Cities in Crisis

Stanisław Kowalkowski · Marek Wrzosek ·
Danuta Kaźmierczak

Cities in Crisis

Challenges for Disaster Management

palgrave
macmillan

Stanisław Kowalkowski
War Studies University
Warsaw, Poland

Marek Wrzosek
War Studies University
Warsaw, Poland

Danuta Kaźmierczak
University of the National Education Commission
Krakow, Poland

ISBN 978-3-032-08842-0 ISBN 978-3-032-08843-7 (eBook)
https://doi.org/10.1007/978-3-032-08843-7

This Palgrave Macmillan imprint is published by the registered company Springer Nature Switzerland AG
The registered company address is: Gewerbestrasse 11, 6330 Cham, Switzerland

Preface

Information society developed high-tech culture and triggered processes which have enabled creation of dynamic, multidimensional and complex environments. The best exemplification of it may be smart city ecosystems, which are perceived as independent, self-regulatory organisms. High-tech solutions led to a merge of cyber, eco and social city spaces that have developed into metropolises functioning as nodes in global networks of trade and communications, business hubs, centres of administrative and political power and sites of cultural production and exchange (Goebel and Keene 2011, Wrzosek 2022).

Succeeding in war requires dominating the places where enemy's politics are conducted and their economic and cultural strength is generated. Thus, consequently cities have become the next domain of warfare (Kilcullen, Weismann. Klaus 2019). Domain of warfare is defined as an operational environment in which armed conflict occurs (McCosker, 2020) requiring specialized tactics, equipment and structure (Dolman, 2022). The urban environment includes the physical aspects of the urban area as well as the complex and dynamic interaction and relationships between its key components—the terrain (natural and man-made), the society and the supporting infrastructure—as an overlapping and interdependent system of systems (FM 3-06). Urban warfare refers to combat in a built environment of some significant size. It is also referred to as Military Operations on Urbanized Terrain (MOUT) or as Fighting in Built Up Areas (FIBUA).

Thus, cities have shaped war but also war has shaped the city form location to street pattern and configuration of public spaces, as well as overt fortification. The inherent complexity of the built-up environment poses some challenge for commanders to apprehend the battlespace, to lead their own forces effectively and to judge the location and intent of enemy forces accurately. Urban warfare is challenging also because of the presence of civilians and sensitive civilian infrastructure (i.e. places of worship, hospitals, museums, etc.) in the battlespace, which limits the choice of tactics and weapons available to commanders for fear of violating laws of armed conflict (Betz 2021). Cities like Aleppo in Syria, Mosul in Iraq and Taiz in Yemen have experienced the deadliest wars of our times as the ICRC reports read:

> People's homes, their doorsteps, their streets, their schools and their hospitals are all part of the battlefield now. Sieges have become all too common, trapping communities between fighters. Devastating explosive weapons terrorize people, young and old. And the violence and losses that people experience day in and day out—in cities that can no longer meet their basic needs—are major causes of displacement within these countries and beyond (International Committee of The Red Cross, 2017).

The conflicts are very often intensified by natural disasters and transformed into humanitarian crises (European Commission, 2021) fuelled by polycrises, i.e. the situations when "economics, politics, geopolitics and then the natural environment [are] blowing back at us. And those four things, they don't reduce to a single common denominator"—vividly explains Adam Tooze (2023, March 7th). These crises are also called complex, compound, protracted and they are heavily impactful especially when war becomes the trigger. Overpopulated, multicultural and networked cities are experiencing all of them.

The scope of the book is defined by the realities of contemporary security environment. These realities have become the focus of our research on cities as a warfare terrain and challenge for civil protection systems. Guided by the logical line along the urban ecosystem as a territory for the critical events and the civilian in their pivotal position who needs the immediate aid provided by disaster risk management systems prepared and organized beforehand (city- urban warfare and crises—DRM system response—Preparation/Education for the future) we formulated the following research questions:

- How the city growths and what are its strengths and vulnerabilities when facing the crisis and war?
- What is the nature of urban warfare and its consequences for citizens?
- How effective the disaster risk management may be in crisis situations?

The methods employed for the research include analysis of the literature, legal acts, reports, statistics and media coverages. We also include case studies to depict vividly and reliably the realities of crises and conflicts.

We believe that the title *Cities in Crisis: Challenges for Disaster Management* we have decided to choose for this book embraces most of phenomena/challenges experienced by city and which we tried to identify and describe: conflicts, wars, warfare, urban combat and crises[1].

[1] For better clarification and precision a selection of exemplary definitions of the terms is provided:

Conflicts, war and competition—"Both war and conflict use a range of means to coerce others: direct organised armed violence; economic and diplomatic actions, trade sanctions, espionage, sabotage, terrorism, insurgency, human security, cyber; and other non-traditional military considerations. What differentiates war and conflict is the primacy of violence. War may be defined as the continuation of a group's; be it a tribal element, community, nation-state or super-state; policy using violence as the primary means of coercion. Conflict is the continuation of a group's policy where violence is one method that either complements another primary means of coercion, or rotates primacy with other non-violent means throughout the conflict. Conversely, competition favours the threat of violence more than the direct use of violence" (Bosio 2020, March 5th).

Warfare- fighting in a war, especially using a particular type of weapon: chemical warfare (Cambridge Dictionary). *Urban warfare* can take many forms, from guerrilla attacks to militia patrols to gang violence, but fighting in cities and towns—where civilians are at risk and combatants can hide more easily—is markedly different from combat in a more open or defined theatre of operations (RAND).

Combat engagement—Action against an adversary, in accordance with IHL/LOAC, NAC-approved rules of engagement, outside of self-defence, to accomplish missions and tasks during operations when it is not feasible to conduct either deliberate or dynamic targeting due to the immediacy of the engagement (AJP-3.9, Allied Joint Doctrine for Joint Targeting)

Urban crises are a multi-hazard reality with higher risk of destruction, loss of assets, and human lives. Therefore, urban crisis response requires capacities from humanitarian actors, which go way beyond their traditional skills European Commission. Humanitarian Action in Urban Crises https://civil-protection-humanitarian-aid.ec.europa.eu/what/humanitarian-aid/urban-crises_en

We look into the city relating to the urban sociology and theory of ecosystem to give the understanding of the complex urban reality in the territorial, cultural and evolutionary dimensions of social life (Scaglia, 2001). The city has been evolving historically into more and more structurally complex settlements: ancient cities, cities in the Middle Ages, modern cities, those of the industrial revolutions and the contemporary, complex, urban ecosystems.

An ecosystem may be defined as "a functional unit consisting of living organisms, their non-living environment and the interactions within and between them" The components included in a given ecosystem and its spatial boundaries are determined by the purpose for which the ecosystem is being created. The boundaries can be relatively sharp or diffused and they can change over time (which happens in case of city boundaries). The city is such a functional "a self-organized social and spatial ecosystem where interactions among (components) people, flows of people, goods and information are only impeded by transaction, transportation, opportunity and matching costs (the cost of matching any one individual's resources, needs, interests and skills with those of another individual) (PLONS). Ecosystems are nested within other ecosystems and their scale can range from very small to the entire biosphere (e.g. local, regional and global cities) (IPCC, 2018). Thus, an urban area includes the city itself, as well as the surrounding areas. City is a densely populated local administrative unit where at least 50% of the population live in high-density clusters/urban centres (Eurostat). Metropolitan area or metropolis is a "large city, including exurbs and suburbs, that has distinct administrative and cultural districts" (Open Education Sociology Dictionary). When two or more metropolitan areas grow until they combine, the area is defined a megalopolis (e.g.: Boston, Massachusetts spread as far south as Washington, D.C. to create the megalopolis of BosWash, or the Northeast Corridor) (National Geographic). Whereas global city demonstrates such "an indifference to the spatial allocation of the production process, supremacy over increasingly broader territories, a market on the world scale and the production of symbols and mechanisms that legitimate the cities' planetary power which the legal and political orders of states are unable to counteract" (Scaglia, 2001). Cities function then as a collective

Urban crises can be defined as "a continuum where chronic vulnerabilities or structural states of crisis can themselves lead to episodic moments of acute shocks which destabilise operations and require urgent responses." (Westman, Patterson, Macrorie et al. 2022).

of complex, networked systems interacting at many scales and as such is discussed in Chapters 1–2.

In the book we use the general terms urban area/terrain or a city with its smart characteristics. The idea of "smart" includes technology as an enabler but a smart city strategy is by far not limited to technological solutions. In fact, "being smart" is more about intelligent methodology and proper implementation of beneficial and effective solutions than about technology. We perceive city as smart ecosystem with its metabolic flows between technical (smart technologies), human resource (smart people) and governance stands (smart collaborations), which enables providing all smart services determining its sustainable development and resilience to shocks, crises and conflicts.

These metabolic flows (supply chains) run through complex, networked and high-tech infrastructure. For its function, the infrastructure has become a critical asset the dwellers are dependent on to the extent that any disruptions in the system make them completely powerless and lead to or intensify the existing crises or conflicts. This human vulnerability and are exploited by variety of adversary groups (states as well as militias). The scarce and essential for life resources (water, energy) and their infrastructure may be a tool or a target in conflicts applied for a political, tactical, or psychological effect. The urban warfare has become the generally accepted warfare concept for today and the future. We analyse it against the following variables: nature of cities (infrastructure, human factor, communication), multidomain operations tailored for *gray zone* competition and processes (globalization, urbanization, technology, size of armies, littoralization).

These variables are interdependent and create very complex operational environment which determine to some degree the nature of urban warfare, the use of weapons and their results. For instance, the attacker has limited cover and concealment, whereas defenders can hide in any locations, use guerrilla tactics to attack and then disappear back into the urban terrain, canalize attacking forces to ambush sites or down roads filled with improvised explosive devices. Technology developments offer a wide range of solutions, equipment and weapons yet, they all may have reduced effectiveness in densely built-up area.

Contemporary urban warfare take a form of multidomain operations—competition in various spheres of the city life rather than a fight to defeat the adversary. It will target the economics (production and allocation), politics (penetration, participation and legitimacy) and cultural identities in the city (e.g.: insurgency/counterinsurgency operation).

Terrorist attacks, problem of cultural inclusion and city affordability in context of influx of uncontrolled mass of migrants will require well designed strategies and far-sighted solutions.

The cases of cities at war in Chapter 3 presents the cities still as the "key elements" of the strategic, operational and tactic level—dictating the final outcome of the combat (STO-MP-MSG-207).

As a consequence of the urban warfare or/and natural disasters the affected population needs both immediate and long term aid which is a challenge for the disaster risk management. To address the needs of the victims, DRM system requires structural flexibility, comprehensive approach and multidimensional and cross sectoral cooperation (following the Sandoval et al. theoretical model of Disaster Risk Management integrated from four interconnected dimensions: sectoral, spatial/hierarchical, temporal and externally with other cross-cutting societal challenges). The case study of the flood in the southern-western Poland (September 2024) in Chapter 4 illustrates how DRM system is able to integrate all necessary entities, especially military units to respond to the crisis. The civil-military cooperation within DRM framework is a bellwether of future character of civil protection (similarly peacekeeping missions resemble very often civil protection/humanitarian operations). Socially sensitive DRM systems accelerate the recovery and resilience building. This trend seems even more justified by the fact that natural disaster may serve as a canvas for soft war tactics: cyber warfare, media warfare and propaganda, civil disobedience, boycotts.

Thus, effective DRM systems need to be based on understanding of disaster risk in all its dimensions of vulnerability, capacity, exposure of persons and assets, hazard characteristics and the environment. This may be achieved through well designed training programs employing the latest teaching-learning theories and LMS solutions. Design thinking and Agile methodology concepts and their practical applications in education are discussed in Chapter 5.

The book attempts to depict the serious global security problems that test our systems of values, governance, war and peace practices and culture as a whole. We believe that it will contribute to building awareness and the public and academic discussion on the present and future condition of cities where most population live and suffer from atrocities of war and prolonged crises. We also hope this book should be a good read/ piece of literature/handbook for academics, policy-makers and practitioners in the field of security, sociology, political studies and disaster risk management.

Warsaw, Poland — Stanisław Kowalkowski
Warsaw, Poland — Marek Wrzosek
Krakow, Poland — Danuta Kaźmierczak

References

Betz, D. (2021, July). Urban warfare. In *Urban Studies*. DOI: https://doi.org/10.1093/obo/9780190922481-0046

Dolman, E. C. (2022). Space is a warfighting domain. *Æther: A Journal Of Strategy And Airpower, 1*(1), 82–90, Spring. https://www.airuniversity.af.edu/

European Comission. European civil protection and humanitarian aid operations. Forgotten crises. https://civil-protection-humanitarian-aid.ec.europa.eu/what/humanitarian-aid/needs-assessment/forgotten-crises_en

Eurostat. Glossary. *City*. https://ec.europa.eu/eurostat/statistics-explained/index.php?title=Glossary:City

FM 3-06 *Urban Operations*. Headquarters Department of the Army Washington, DC, 26 October 2006.

Friedman, J. (2005). Globalization and the emerging culture of planning. *Progress in Planning, 64*(3), 183–234. https://doi.org/10.1016/j.progress.2005.05.001

International Committee of The Red Cross. (2017, June 14). *I saw my city die: Voices from the front lines of urban conflict in Iraq, Syria and Yemen. Special Report*. https://www.icrc.org/en/publication/i-saw-my-city-die-voices-front-lines-urban-conflict-iraq-syria-and-yemen

IPCC. (2018). Annex I: Glossary. In J. B. R. Matthews (Ed.), *Global Warming of 1.5°C. An IPCC Special Report on the impacts of global warming of 1.5°C above pre-industrial levels and related global greenhouse gas emissionpathways, in the context of strengthening the global response to the threat of climate change, sustainable development, and efforts to eradicate poverty* [Masson-Delmotte, V., P. Zhai, H.-O. Pörtner, D. Roberts, J. Skea, P.R. Shukla, A. Pirani, W. Moufouma-Okia, C. Péan, R. Pidcock, S. Connors, J.B.R. Matthews, Y. Chen, X. Zhou, M.I. Gomis, E. Lonnoy, T. Maycock, M. Tignor, and T. Waterfield (eds.)]. Cambridge University Press, Cambridge, UK and New York, NY, USA, pp. 541–562. https://doi.org/10.1017/9781009157940.008.

Klaus, I. (2019, August 01). Intuition, the City, and War. https://mwi.westpoint.edu/intuition-city-war/

M 3-06 Urban Operations. Headquarters Department of the Army Washington, DC, 26 October 2006

National Geographic. *Urban Area*. https://education.nationalgeographic.org/resource/urban-area/

RAND. *Urban warfare*. https://www.rand.org/topics/urban-warfare.html)

Scaglia, A. (2001). Urban sociology. In *International Encyclopedia of the Social & Behavioral Sciences* (pp. 16082–16087). Science Direct. https://doi.org/10.1016/B0-08-043076-7/02029-5

STO-MP-MSG-207.Smith M. and Robson, D. *A Concept for the Modelling and Simulation of Complex Urban Environments*. NATO DSTL/CP152445

Tooze, A. (2023, March 7). Where did the term polycrisis come from and what does it mean?. WEF. https://www.weforum.org/stories/2023/03/polycrisis-adam-tooze-historian-explains/

Westman, L., Patterson, J., Macrorie, R., et al. (2022). Compound urban crises. *Ambio, 51*, 1402–1415. https://doi.org/10.1007/s13280-021-01697-6

Competing Interests The authors have no competing interests to declare that are relevant to the content of this manuscript.

Contents

List of Figures

List of Tables

CHAPTER 1

City—A Highly Complex Ecosystem to Live (Operate) in

Abstract Urbanization is a multidimensional process that manifests itself through rapidly changing human population and changing land cover. The combination of multifold driving forces physical, geographical and human determine city's individual dynamics and structure which translate into the city models. In this chapter city is described as a smart ecosystem—a human centric, highly complex, digitalized, networked construct. It discusses the city life now, reflecting on the past and post-global future.

Keywords Models of the city · Smart city · City dynamics · Post-global city

1.1 City—Definitions

The origin and development of cities was fostered by three major transformations: the revolution that led to the development of agriculture around 7000 BC and Neolithic farming settlements (Jarmo, Iraq and Jericho, Israel); the pre-industrial revolution that brought cities into being; the industrial revolution of the eighteenth and nineteenth centuries that created the urban industrial harbingers of the present cities.

S. Kowalkowski et al., *Cities in Crisis*,
https://doi.org/10.1007/978-3-032-08843-7_1

A crucial element to understand cities is the process describing their growth—urbanization.

Urbanization is the process through which cities grow and higher and higher percentages of the population come to live in the city (National Geographic). The rise of early civilizations and integration in the design of cities were fostered by three factors: (1) mobility of people to available resources, (2) ecosystem management to secure enhanced local growth of produce and (3) increasing social complexity encoded in formal institutions that guided an expanding range of activities.

The growth of cities intensified in pre-industrial and industrial societies as the result of the following processes:

- the industrial revolution and consequently rural emigration;
- the rise of the tertiary sector;
- the need of many households for a broad employment market, taking into account the growing diversity of professions and the accompanying demand of professional mobility, both desired and constrained;
- the rise of "world space", which often creates the necessity of sophisticated and highly functional international connections
- the political strategy exercised in matters of implantation and structuring of the territory,
- the policy regarding fostering of migration and specific urban government planning (Dumont, 2013).

The urbanization processes today are different from urban transformations of the past in terms of (1) the scale, (2) the rate and (3) the shifting geography of urbanization. It may be seen as a multidimensional process that manifests itself through rapidly changing human population and changing land cover. These changes result from a combination of four driving forces: natural growth, rural to urban migration, massive migration due to extreme events (Springer p. 4) (conflicts, natural disasters, economic or political unrests) and redefinitions of administrative boundaries.

Urbanization is a process and the dichotomy urban-rural does not describes the dynamic nature of human settlements and has limited comparability internationally. For that purpose, the *Degree of Urbanization* (DEGURBA) is applied. It identifies three types of settlements capturing the urban-rural continuum (UN HABITAT):

1. Cities, which have a population of at least 50,000 inhabitants in contiguous dense grid cells (>1500 inhabitants per km^2);
2. Towns and semi-dense areas, which have a population of at least 5000 inhabitants in contiguous grid cells with a density of at least 300 inhabitants per km^2; and
3. Rural areas, which consist mostly of low-density grid cells (<300 inhabitants per km^2).

Urban extent is another tool. It defines city by its urban extent (built - up and urbanized open space). Urban extent represents the total built-up area, which is itself defined as the contiguous area occupied by buildings and other impervious surfaces. The urbanized open space on the other hand refers to unbuilt-up areas which are encompassed within the built-up areas or within their immediate vicinity and include parks, cleared land, forests among others.

Cities earned also sociological definitions. A city is a relatively large, dense and permanent settlement of heterogenous individuals (Wirth, 1938) who differ from individuals inhabiting villages. Their distinctive quality is sociologically defined as a distinctive lifestyle (Siebel, 2018). City is not merely a physical/geographical place (locale), but space (milieu) of threefold processes:

- frequent and intense social interactions among a diversity of individuals and organizations like trading and all the activities of human lives (sleeping, working, eating, playing);
- cultural reflecting values, attitudes, ways of thinking and acting;
- political shaping relations of power and control (Jayaram, 2010).

This functional perspective leads to the definition of a city as" a self-organized social and spatial arrangement where interactions among people, flows of people, goods and information are only impeded by transaction, transportation, opportunity and matching costs (the cost of matching any one individual's resources, needs, interests and skills with those of another individual) (Ortman et al., 2020, p. 4/24). Cities function then as a collective of complexes, networked systems interacting at many scales (local, regional and global). They continue to grow as major/world centres of population, economic development and innovation and are often seen as both the cause and solution for many global environmental problems (Sparks et al., 2020).

Method: Relating to the urban sociology and the theory of ecosystem we present different evolutionary structures and organization of urban spaces (models of the city) in different regions of the world, which create a varied terrain for the urban combat discussed in Chapter 3. The dynamics of the smart city ecosystem is discussed from the human-centric perspective in the following areas: governance, economy smart living, mobility and people (the indicators adapted from Smart City Index 2019). The discussion is completed with some reminiscence of pre-war Warsaw (the exemplary historical picture of a big city) and speculations about post-global cities to embrace the full-time line. Human centric perspective is continued in Chapter 2 dealing with the malfunctions of the city infrastructure and their impact on city-dwellers in time of shocks and prolonged crises.

1.2 Models of the City (Cities in Different Regions of the World)

Urban models are based on the geometry of the spatial structure and the assumption that social processes generate these spatial forms. It is then possible to create a synthetic vision of society-space relationship in the particular area (Buzai). City may be viewed through the "social disorganization" as its organizing principle and interpretive framework. The city described as "a mosaic of social worlds" (Park & Burgess 1925/1967), residentially and racially "hyper segregated" (Denton, 1994), or "a non-contiguous collage of parcelized, consumption-oriented landscape" (Dear & Flusty, 1998), it is simply fractured and fragmented (Borer, 2006). Other analytical parameters of urban structure applied are: functions (residential, commercial, industrial); socioeconomic characteristics of residents; density of development; locations of activities and interrelations. Urban models prove that the structuring of cities is influenced by their history, economy, politics and social-functional mix. These factors interact with the conventional parameters to produce new forms of urban structures, which create the spectrum of structural models ranging

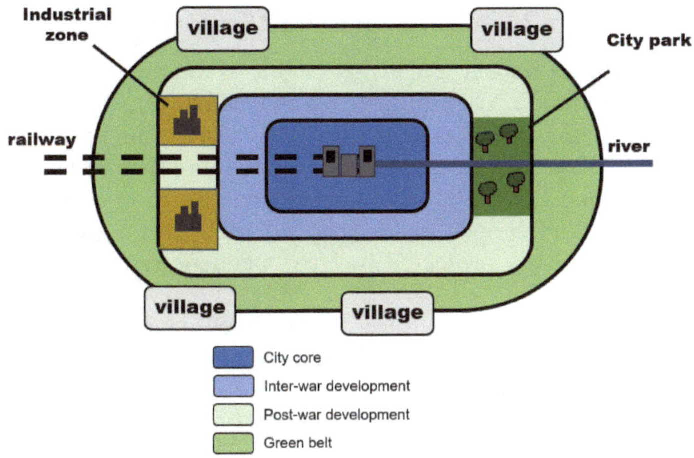

Fig. 1.1 European cities model (*Source* based on: Joseph Park on: Mon 26 of April, 2010 00:03 CDT (56,165 Reads); Park and Burgess (1925), The City)

from centred to decentred, nucleated to networked functional districts (Qadeer, 2024) (Fig. 1.1).

1.2.1 Hopkinson's Model, 1985

Hopkinson's model constructed to reflect the structure of the UK and Western European cities. It incorporates both concentric zones and sectors. The earliest factories created a ring around the old city (now the city core) whereas the recent industry is located along railway lines. The residential areas extended from the central business district (CBD) along the main roads and in post-war period spread in-between the inter-wall residential areas.

The CBD is consolidated on an historic, usually defensive site such as an old bronze age hill fort and/or a mediaeval castle) and remains the historic and cultural heart of the city and the administrative centre.

1.2.2 Latin America City Model

The model shows a central sector that begins to fragment, a periphery of undefined boundaries. According to Janoschka (2002), the new peripheral location of the urban elite is providing new forms to cities in the XXI century. The gated communities are much more than an architectural phenomenon, being, mainly, an important manifestation of social polarization. The cities with open public spaces demonstrate a greater number of enclaves increasingly protected by bars, magnetic cards and personnel security (Fig. 1.2, 1.3).

The modern city in the Southeast Asia is well depicted by the model by T. G. McGee. His land-use model illustrates the locations of old (the port zone) and new (the industrial estate) economic activities as well as residential and commercial zones which existed during colonial times and those created recently. The new peripheral suburbs include the low-income squatter areas so prevalent in Southeast Asian cities today. The

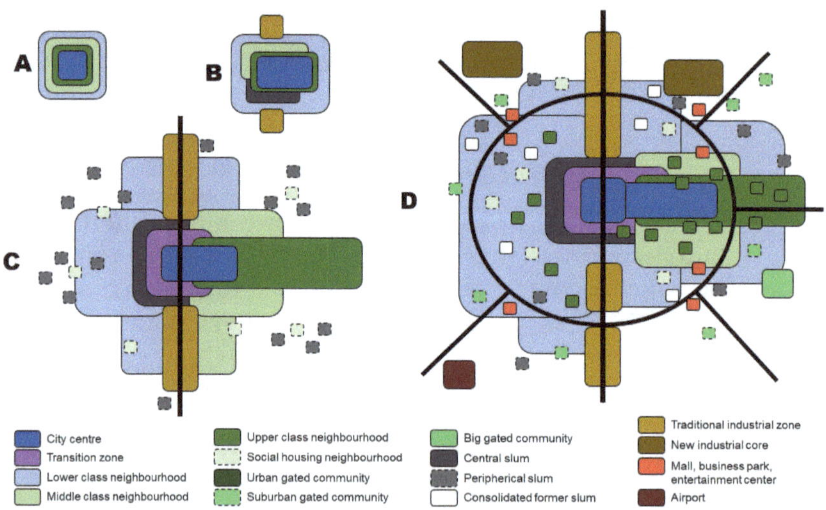

Fig. 1.2 Latin America City Model (*Source* based on: Hoyt, H. (1939), The structure and growth of residential neighbourhoods in American cities (Washington, DC))

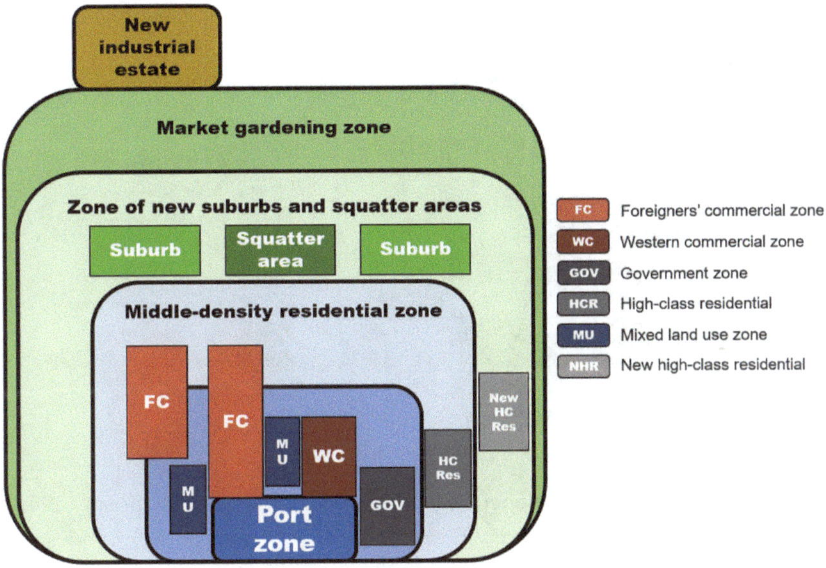

Fig. 1.3 The modern city in the Southeast Asia (*Source* based on: Janoschka (2002))

commercial zone, located in the CBD, generally has the highest population density and is inhabited by the Chinese residents (Moniruzzaman et al., 2013).

1.2.3 Multiple Nuclei Model

Multiple-Nuclei model of 1945 by C.D. Harris and Edward L. Ullman is a model of urban land use in which a city grows from several independent nodes rather than from one central business district. Each node acts as a centre for a particular kind of land use, such as industry, retail, or high-quality housing. As these expand, they merge to form a single urban area. The CBD is not the only driver of development. Other nodes include: ports, universities, airports, parks, neighbourhood business centres. This model indicates the varied factors of decentralization in the structure of the North American city, for example Los Angeles. The multiple-nuclei theory based on the idea of people mobility due to increased car ownership as a trigger for the specialized regional centres

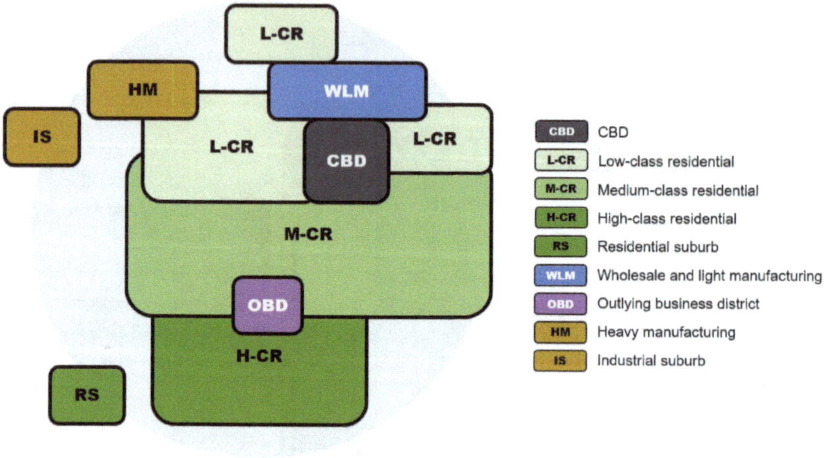

Fig. 1.4 Multiple Nuclei Model (*Source* based on: Harris and Ullman (1945), The Nature of Cities. *Annals of the American Academy of Political and Social Science*, 242/Nov.: 7–17)

(e.g. heavy industry zones, business park). Multiple nuclei models are complex, allows the suburbanization, e.g. industrial suburbs, with transport development and outward growth of city. He multiple nuclei model is criticized for negligence of height of buildings, non-clearly marked divisions between zones which demonstrate a significant degree of internal heterogeneity and not homogeneity. Moreover, this concept may not be applicable to oriental cities with different cultural, economic and political backgrounds (Fig. 1.4).

1.2.4 Galactic City Model

Lewis (1983) coined the term "galactic metropolis" to capture disjoined and decentralized urban landscape of the twentieth century North America: "Galactic metropolis is vast with varying sized urban centres, subcentres and satellites. It is fragmented and multimodal, mixed densities and unexpected juxtapositions of form and function" (Neuman & Hull, 2011, p.6). Hall (2001) identified the following nodes characteristic for polycentric metropolis which are demonstrated even by older, established metropolises, for example Detroit (Neuman & Hull, 2011) (Fig. 1.5):

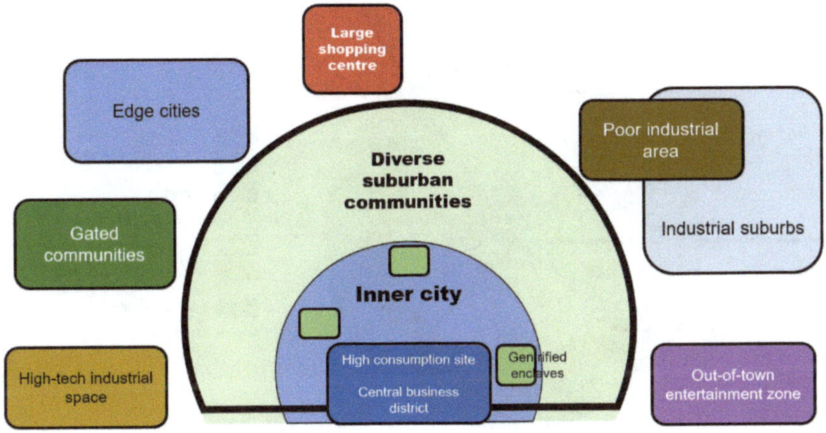

Fig. 1.5 Galactic city model (*Source* based on: Lewis (1983), Hall (2001))

- traditional downtown centre—the hub of traditional metropolis, now location of informational services: government, banking and insurance;
- newer business centre developing in prestigious residential quarter as a location of business, media, advertising, public relations;
- internal edge cities—developing as a result of pressure for space in traditional residential areas and speculative development in industrial sites,
- external edge cities—developing on the axis of the major airport, linked to the urban highway system,
- outermost edge city complexes—for the research and development centres near major transport hubs,
- specialized subcentres for education, entertainment, sporting and exhibition complexes.

1.2.5 African City Model

African cities are influenced by European colonialization. The difference between African and Latin American city structures is that African cities have three CBDs: a traditional open market, a European colonial centre with grid-like streets and a developing CBD. These CBDs are placed in

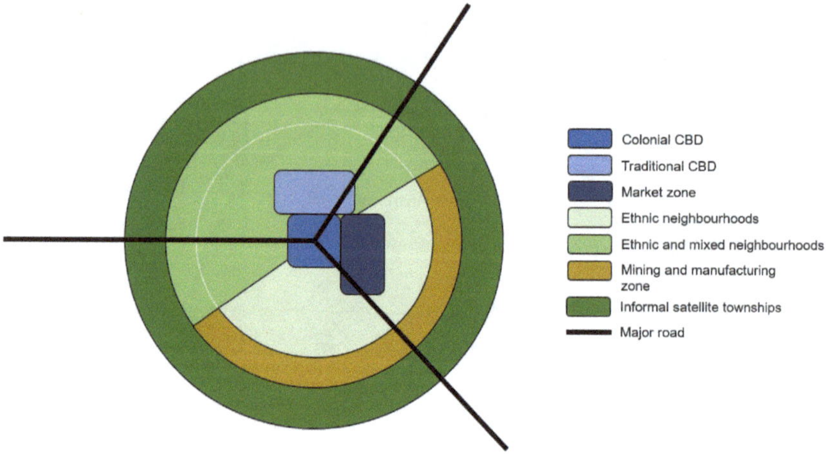

Fig. 1.6 African city model (*Source* based on: de Blij Subsaharan Africa city model (based on Fouberg et al., 1977)

the centre with residential areas surrounding them. These residential areas differ depending on the proximity to the CBD. The residents closer to the CBD vary in socio-economic terms. Manufacturing zones surround these residential zones. After the manufacturing zones the squatter settlements spread along the outskirts of the cities (Josse, 2020).

Over the past century urban regions have been evolving from centralized settlements to polycentric structures, to ultimately form networked megaregions which function as service- and knowledge-based economies fostered by advancements in infrastructure and technology. The urban structure encodes not only historical, cultural, economic evolutionary patterns but also ethnical ones. Ethno-racial diversity permeates the structure of the city structures, creating distinct patterns of geographic distribution of racial and cultural groups, their institutions and their symbols (Fig. 1.6).

1.3 Smart City Ecosystem

Evolution and expansion of cities resulted in their interlinked social and technological systems diversity transforming them into complex urban ecosystems.

Jane Jacobs defines them as "organized complexity" which means "dealing simultaneously with a sizeable number of factors which are interrelated into an organic whole" [Jacobs, p. 432]. Cities share the general properties of a complex system in a way that they are non-extensive and their densities are non-intensive; that is their overall properties are not merely a sum of their parts and they also depend on the size of the entire system. Cities' spatial and individual heterogeneity is demonstrated in a way that:

> [...] there are rich and poor neighbourhoods, there are business districts, which are almost exclusively dedicated to jobs and trade and there are suburbs, which are mostly residential. There are parts of the city dedicated to certain types of business, manufacturing, etc. Likewise, there are poor and rich individuals, people with radically different economic and social expertise, motivations, ethnicity, etc. The city only makes sense when all these parts are put together as an interacting social system (Bettencourt et al., 2013, p. 17).

It is not possible to distil a representative average place or average person inside the city. Analysing cities as complex systems requires examining settlements across geographies and eras to identify commonalities, observe variations and determine how the two are related (Ortman et al., 2020).

Complexity of the city results not only from socio -economic interactions but also from the human- Earth dependency—an approach typical of Anthropocene (Alberti, 2024). This emphasis results in the cities being viewed as complex social-ecological systems (Alfsen et al., 2011) in a manner that:

> Urban megaregions, characterized by high population densities, resource concentration and central roles in trade and logistics, can either exacerbate or mitigate the environmental footprint of global production and consumption systems. These regions accelerate eco-evolutionary changes by increasing resource demand, modifying land management and facilitating the translocation of non-native species, extending their impacts beyond their physical boundaries (Alberti, 2024, p.3).

The ecological variations caused by unequal distributions of resources and access to ecosystem services among socio-economically diverse communities develop the eco-evolutionary dynamic urban ecosystem sectioned off other natural and anthropogenic systems.

This characteristic urban spatial structure results from the interactions of human agency, infrastructure and the physical environment.

David Kilcullen (2015) refers to the "urban social metabolism" model while discussing cities' stability, sustainability and resilience. This model explains how the city transforms the inputs, manages by-products of that transformation and how this city's ecosystem nests within the national and global systems. Material flows include food, air, water, electrical power and fuel), economic flows—construction materials, licit and illicit commodities, air traffic, ground, sea, money. Along with material, economic and information flows the population flows contribute to creation of peri urban settlements (slums, squatters, shantytowns) around the old city core (CBD). The city's growth strains its infrastructure and governance—the "carrying capacity" to metabolize inputs. This situation leads to toxic often cascading and compound risks: urban poverty and exclusion, disease, unemployment, social injustice and ethnic dislocation generate violent crime, social and political unrest and even organized conflicts. This violent urban ecosystem is facilitated by shortages of food, fuel, electricity and water. Whereas city connectedness via information and money flows and transportation hubs influences population's activity in the rural hinterland (local and national level) and global networks. This set of interactions fuels local and international conflicts.

1.3.1 Smart Technology

Each city is created in a particular environment with individual topographical and fertility characteristics. As time and technologies develop this relationship becomes less dominant. Technology and governance are the "factor condition" at play, which impact city economies, culture and built forms. Changes in technology lead to new products, new production processes, changes in a skills base and division of labour, new architectural forms and culture with digital media.

Urban social, economic, cultural and political life is supported by background technological networks and systems (infrastructure):

> With infrastructure, the central dynamic is around the facilitation of activity. That could well be about the facilitation of water distribution, sanitation, electrical power, or communication through hard technological systems, but it can also be about much more than that (Farkas et al., 2009).

Ever- expanding networks (connectivity) are neither licit nor illicit, they cannot be judged, but behaviours within them can. The crucial characteristic of networks is their invisibility. D. Kilcullen calls them dark as they resemble "subterranean rivers of connectivity that runs below and between the elements of the world we see. They cannot be observed directly unless we do soothing to stimulate network, drawing a detectable response" (Kilcullen, 2015, p. 34). They engage in all sort of activities, beneficial, neutral and harmful, all at the same time.

Social networks for example offer dwellers variety of lifestyles: family reunions and kitty parties, clubs and associations, cult groups and secret societies, chit funds and mutual-aid groups and so on: "One could postulate that the stronger one's social networks in the city, the more comfortable one would be her or his life; conversely, urban life would be wretched without social networks" (Jayaram, 2017, p. 26).

Understanding the presence of these networks and their multipurpose nature and the way their flows intersects is crucial for our understanding of the future environment.

The interplay of socio-economic elements, infrastructure and technology create the backbone for variety of concepts for city development: digital city, smart city, cyber city or smart cyber city.

Digital city simulates real cities by computer science and geographic science to help government for city planning and transportation simulation and to implement city-wide network infrastructure and public/private services upon it. Digital city may be viewed also as a cantered on a city website, where public authorities, business and citizens communicate and exchange information. E-government and e-governance is to build the relationship among citizens and public, business and nonprofit sector. There are some famous digital cities including Virtual Los Angeles, Model City Philadelphia, Google earth and virtual earth (Cocchia, 2014; Farkas et al., 2009; Jaeger, 2009; Zhu, 2009).

Cybercity, which is another concept of urban development, characterizes with the complex interlinkages between telecommunications, urban form and transportation (Graham & Marvil, 1999). Cybercities will foster the greater economic productivity, social cohesion, resilience and environmental sustainability of the city entering into the fifth wave of wealth creation built on technology and new industries. Although they will not be free from the depressing images of slums, they will still offer more opportunity to more people at less social and environmental cost than ever (Mead, 2024; Montgomery, 2008).

This human-centric (or Kilcullen's urban social metabolic) approach is ingrained in the concept of smart city which is based on the integration and the networked flows between technical (smart technologies), human resource (smart people) and governance stands (smart collaborations), which enables providing all smart services: smart living, smart transport, smart production, smart energy, smart governance and as a result, a smart planet. "Smartness" is directly related to "awareness", which is the ability or state to recognize, to sense, or to be conscious of surrounding objects, events, or physical configuration. Being smart is not an exclusive attribute for cities. Every community—regardless of size or geographical structure (urban, rural communities)—can be smart in the way it responds to challenges and pursues opportunities for community transformation. Another factor that is of high relevance in this respect is the connection between urban cities and rural communities. Often, they fulfil different but complementary functions in a system or network of central places (the concept of centrality). Thus, smart solutions and smart city projects should take the connection of urban and rural communities into account. Smartness results in sustainable and efficient city driven by whole spectrum of technologies: Artificial Intelligence, Internet of Things, Cloud Computing, Multi Agent Systems, Sensors, Metaheuristics. These technologies create the specific operational environment of the smart city, thus city itself can be viewed as—a Cyber Physical System—i.e. complex systems composed of organic integration and in-depth collaboration of computation, communications and control (3C) technology—smart cyber city. Yet, the term "smart" includes technology as an enabler but a smart city strategy is by far not limited to technological solutions. In fact, "being smart" is more about intelligent methodology and proper implementation of beneficial and effective solutions than about technology. A smart city integrates technology to accelerate, facilitate and transform city ecosystem—comprised of people, organizations and businesses, policies, laws and processes coordinated to create the desired outcomes (Fig. 1.7).

The ecosystem of implemented solutions is what defines how smart a city/community is. Yet, these solutions need to be less technology-centric

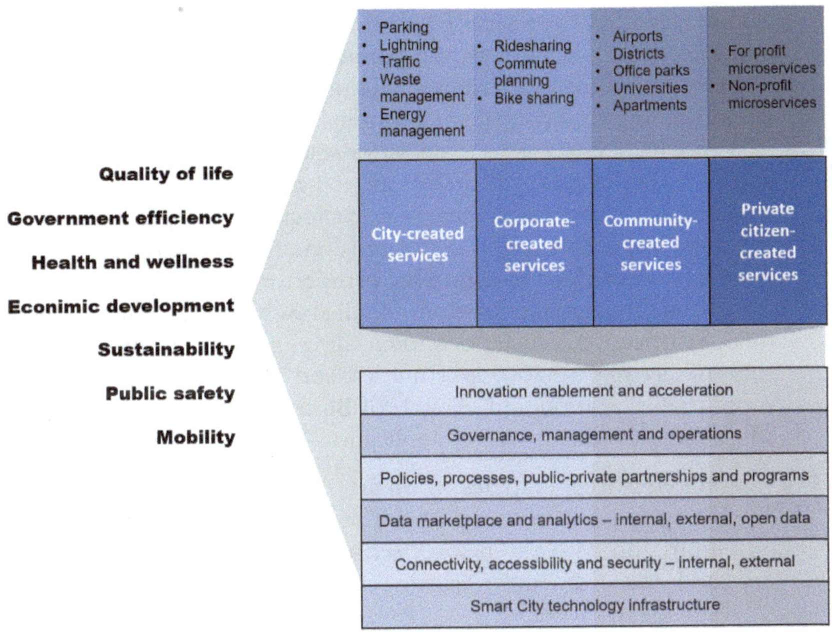

Fig. 1.7 Smart city ecosystem (*Source* Based on: IIoT World, The Smart City Ecosystem Framework—A Model for Planning Smart Cities, January 26, 2018; https://www.iiot-world.com/smart-cities-buildings-infrastructure/smart-cities/the-smart-city-ecosystem-framework-a-model-for-planning-smart-cities/ (2.10.2024))

and more human-focused (Smart City Index in 2019). The indicators for smartness may be as follows: smart government, smart economy, smart people, smart mobility, smart environment, smart living.

1. "**Smart Government**" is about strengthening the connections and interactions between the government and all stakeholders–citizens, businesses and other organizations of the civil society—within a municipality. By utilizing new methodologies, such as co-creation or crowd-sourcing, or by implementing new technology and innovation (e.g. for digital citizen or business services or the management of public infrastructure) a "smart government" can be developed.

Following a "city as a service" model can help to increase efficiency and effectiveness as well as transparency and trust.

2. "**Smart Economy**" describes all actions aimed at transforming and strengthening a municipality's economy. Improving the overall business climate, a city's attractiveness for start-ups, investors, businesses and new (highly qualified) talent as well as growing the economy in an innovative and sustainable way to increase competitiveness are the most important goals. Utilizing (digital) technology and intelligent approaches lead to economic prosperity that, in turn, generates stable and favourable conditions for all stakeholders. From a government perspective, "smart economic development" is an important tool to actively seize opportunities and provide conditions that support the creation and growth of businesses as well as new jobs. The examples of technology application in that area are:

   - Wireless Technologies in Smart Agriculture Wireless connectivity can enhance and optimize conventional farming operations and with the help of sensors, robots, computer imaging and analytical tools, farmers can gain control over their livestock and crops.
   - Greenhouse Automation Accurate real-time information on greenhouse conditions such as soil condition, lighting, temperature and humidity.
   - Cattle Management and Monitoring Sensors attached to cattle to monitor their health and identify sick animals to separate them from the herd and avoid contamination.
   - Precision Farming Based on sensor data, farmers get to estimate the optimal amounts of water, fertilizers and pesticides that their crops need, reduce expenses and raise better and healthier crops.

3. "**Smart Environment**" The reduction of waste production, monitoring and managing pollution, emission reduction, water management, achieving energy efficiency and accelerating the local energy transition are some important goals of "smart environment" initiatives. New urban planning standards to improve efficiency and to minimize the environmental impact, as well as the creation of a resilient community are further goals.
4. "**Smart Living**" aims at increasing quality of life for residents and visitors by following an inclusive strategic approach—across all age

groups and demographics. Smart Living focuses on improving social and digital inclusion (e.g. the use of electronic services, connectivity and social platforms), on improving healthcare and care for the elderly (e.g. eHealth, Ambient Assisted Living), safety, housing conditions and smart buildings. New methodologies for civic and social engagement as well as new technologies (e.g. IoT based on Wi-Fi or LPWA network technology) are leveraged to improve accessibility and citizen experience across all focus areas. LPWA technologies have been developed for IoT products to enable wireless communication at low-power consumption but long-range. As the Internet of Things technology evolves, low-power consumption for electronic devices and products is increasingly important

5. "**Smart Mobility**" focuses on increasing the efficiency and service quality of urban transportation. Achieving cheaper, faster and environmentally friendly mobility as well as integrated multi-modal transportation is an important challenge for cities and communities. Supporting the combination of multiple modes of public and private transport and adopting new forms of transportation (e.g. electric vehicles, hydrogen-powered vehicles, autonomous vehicles, bike sharing, carpooling/car-sharing) is an important aspect for a future-oriented strategic approach to foster "Smart Mobility". To achieve a high-quality mobility service and to ultimately improve the flow of people and goods within a city or community, while at the same time reducing the environmental impact.
6. "**Smart People**" First, "Smart People" aims at transforming the way citizens interact—via information or the provision of services—with the public and private sector as individuals or businesses. Creating social and digital inclusion/digital equality through educational offers is an important prerequisite for a more efficient provision of information and services based on new technologies. Second, "Smart People" is about smart forms of education to facilitate career choices, labour market opportunities, vocational training as well as lifelong learning for all age groups and demographics. Talent development is also an important aspect from an economic development perspective as an increasingly important location factor.

Although the cities have been evolving and the methodologies too, London, New York and Paris still top the rankings for smart and sustainable cities. In the most recent report, IESE analyse 183 cities in 92 countries against the following indicators (Cities in Motion Index 2024):

- **Human capital**—the education level of the population and their access to culture.
- **Social cohesion**—protections for vulnerable groups, community development and healthcare.
- **Economy**—current and estimated GDP, innovation activities and, the number of unicorn companies and companies on the Fortune Global 500 list.
- **Governance**—the quality of state interventions.
- **Environment**—pollution levels, water quality and other gauges of ecosystem vitality and climate vulnerability.
- **Mobility and transportation**—Ease of movement and access to public services; bicycle ownership rates and the length of metro lines.
- **Urban planning**—health infrastructure, sanitation services and housing policy, artificial intelligence projects that promote urban growth and wellbeing.
- **International profile**—Branding and strategic tourism outreach (hosting international conferences, hotel offerings and the Restaurant Price Index).
- **Technology**—Smart uses of ICT technologies, broadband availability and the percentages of households with computers/mobile phones.

They have identified the top 10 cities in the world as follows:

1. London, UK
2. New York, USA
3. Paris, France
4. Tokyo, Japan
5. Berlin, Germany
6. Singapore, Singapore
7. Oslo, Norway
8. Amsterdam, The Netherlands
9. San Francisco, USA

10. Chicago, USA

Warsaw, Poland takes the 54th rank. From Latin America, the highest-ranked city is Santiago (91st), Buenos Aires (115th) and Montevideo (119th). Cape Town (141st) is the top-scoring city in Africa. Sydney, Australia occupies the 28th position, Seoul, South Korea—the 13th and Benjin, China—the 21st.

1.4 Cities in Post-Globalization Era

In 2023, around 56% of the world is urbanized by 2045, the world's urban population will increase by 1.5 times to 6 billion. Cities generate more than 80% of global GDP. The evolutionary route of cities has been paved by the coupled processes of technological and cultural development, economic growth and globalization (Friedmann, 1986). Moreover, the city dynamics is influenced by external factors modified by endogenous conditions. The city's integration with the world economy depends of these two forces. The endogenous condition is shaped by the spatial patterns of historical accumulation, national policies which protect the national economic sub-system from outside competition through partial closure to immigration, commodity imports and the operation of international capital as well as social conditions. Whereas globalization processes are external forces influencing these internal conditions and consequently, allow to analyse the city in economic terms (city as a production site) or social and cultural terms (urban living).

City viewed from a global perspective, may earn the position in a global system of an economic command centre/agency or an object admitting the products of globalization. This criterion made it possible to arrange world cities into a complex spatial hierarchy: the first level: 3 primary cities that dominate global economy: NY, London, Tokyo; the second level: Paris, Los Angeles; the third level occupies the cities which role is confined to the regions they are located in (Miami, Singapore) (Friedmann, 1986).

Leading cities have been growing into megacities—an urban agglomeration with 10 million inhabitants or more (UN, 2018). In 1950 the world counted just two—New York and Tokyo—with only Mexico City joining their ranks in the following 25 years. Today, there are 40 megacities around the world and that number is expected to surge in the years to come. Nine of the 10 cities projected to become megacities between 2018 and 2030 are located in developing countries (Table 1.1) as more developed areas are to experience stabilization or decline. Yet the world will be more and more urbanized.

The urbanization will develop along the following pattern:

- Cities around the world will grow, into "mega-global cities" will offer more powerful and advanced business services sectors or international financial centres and "mega-local cities" that will not;
- A particular country or territory will potentially have one megacity for every 100 million people in the country;
- "Internal" forces will be at work: the prime city of each country will dominate their respective urban structures and occupy an increasingly-larger share of national population and determine the growth of the largest cities;

Table 1.1 The world largest cities in 2018 and 2030

The world's ten largest cities in 2018 and 2030

City size rank	*City*	*Population in 2018 (thousands)*	*City*	*Population in 2030 (thousands)*
1	Tokyo, Japan	37,468	Delhi, India	38,939
2	Delhi, India	28,514	Tokyo, Japan	36,574
3	Shanghai, China	25,582	Shanghai, China	32,869
4	Sao Paulo, Brazil	21,650	Dhaka, Bangladesh	28,076
5	Mexico City, Mexico	21,581	Al.-Qahirah (Cairo), Egypt	25,517
6	Al.-Qahirah (Cairo), Egypt	20,076	Mumbai (Bombay), India	24,572
7	Mumbai (Bombay), India	19,980	Beijing, China	24,282
8	Beijing, China	19,618	Mexico City, Mexico	24,111
9	Dhaka, Bangladesh	19,578	Sao Paulo, Brazil	23,824
10	Kinki M.M.A. (Osaka), Japan	19,281	Kinshasa, Democratic Republic of the Congo	21,914

Source UN (2018). The World's Cities in 2018

- “External” forces: IT and finance and business sectors will be the key “external” factors driving the growth of cities and megacities, transforming “mega-local cities” to “mega-global cities”;
- Technologically advanced transport and telecommunication networks are other external factors driving megacity growth (Zhao et al., 2017).

These predictions could have been unquestionable in globalization era, however COVID-19 pandemic forced us to revised them slightly. Although the future of humanity is undoubtedly urban, it will be shaped in different environments and contexts. Zeihan (2022) draws the gloomy future of the globe: the collapse of long-range transport and intercontinental supply chains will have the impact on access to energy and industrial materials. These factors accompanied with negative demographics in most countries: in China, Russia, in the European Union (except the US with a less declining birth rate trend) will cause the world to face reduced consumption, slower economic growth and consequently less capital generation. We will lack common products and production or import will become more expensive; and hunger will spread over the planet. The epoch we are entering is not merely ‘de-globalization’, or even ‘de-industrialization’, but ‘de-civilization’—cascade of collapses with feedback loops which not only break but destroy the operational foundations of the present world. The cities especially Asian megacities will be affected the most. The reduced global system and transport will force them to provide for food, energy and factors of production (land, labour, capital and entrepreneurship) on their own. Only cities in the block will be able to supply for dwellers food, heating and offer employment. Most world urban population will share the same fate—mass deindustrialization and migration to the villages (Zeihan, 2022). It is estimated also that African cities could lose up to two-thirds of their financial resources and the urban service delivery and governance system could collapse (UN, 2022). Another only slightly less depressing one is a “business as usual” scenario in which cities will attempt to return to the pre-pandemic state of affairs and as a result they will deepen disparities, exclusion of the underprivileged, excessive exploitation of natural resources, digital inequalities and limited health care service—the environment prone to variety of vulnerabilities. The favourable development could be ensured only by adopting sustainable approaches—collaborative, coordinated multilateral interventions to leverage the opportunities and mitigate the impacts of crises (UN, 2022). The digital city, smart cyber city, or smart city concepts presented above are the approaches that aim at building urban resilience—the inherent property of a social-ecological system that enables it to maintain its essential functions in the face of dynamic changes (Alberti, 2024).

1.5 Warsaw Before WW II—Reflection on Smart Development

Warsaw, the capital of Poland may serve as the example of the city that developed very fast exploiting all the economic, cultural, political opportunities brought by the times before the WW II. This is also the city that in spite of being completely destroyed during WW II, has recovered and is developing to earn the 38 positions in IMD Smart City Index 2024 (in 2023 Warsaw has 44 rank)[1] which assesses cities in two pillars for which perceptions from residents are solicited. The Structures pillar refers to the existing infrastructure of the cities and the Technology pillar describes the technological provisions and services available to the inhabitants. Each pillar is evaluated over five key areas: health and safety, mobility, activities, opportunities and governance.

Although there are not the scores of Smart City Index of the period before WW II available, we have made an attempt to picture the life in Warsaw at that time and we retrieved some data from the archives. The statistics were published by Statistic Department of City Management Board in 1938 (Wielka Historia). We analysed and grouped the information into five Index areas: health and safety, mobility, activities, opportunities and governance. The results are as follows:

1. Health and Safety

Road infrastructure. In the late 1920s Warsaw had 1800 streets with names and marked out in the field, in total they were 780 km long but only 485 km (62%) were cobbled. Today Warsaw has over 5000 streets, alleys, squares which in total measure 3000 km.

Street lightning. There were 87 per cent streets lightened at night. However, the statistics in Warsaw in figures Booklet informs that 510 km

[1] According to the quoted above Cities in Motion Index 2024 Warsaw took 54th place.

of roads were lightened, i.e. 65% of all street length. 35% of the city, especially peripheries were shrouded in darkness. Warsaw was lit by 15,228 lampposts including 8686 electric ones and 6542 gas lamps.

Fires and arsons. In 1937 Warsaw experienced 681 fires and 237 of them were caused by the faulty constructions and 197—carelessness. 9 explosions and over 200 arsons for unknown reasons were reported. Only 26 false alarms were reported. In 250 cases the property owners received the damages from the insurance companies (in total 156,000 zlotych which equals today 1,500,000 PLN).

Access to water, electricity and gas.

The water infrastructure supplied water to 90 per cent of citizens, however 70 per cent had access to sewage system. Only 800 buildings (mainly private houses has access to gas infrastructure). Gas consumption reached 70 per cent for cooking and 30 per cent for industry. Citizens enjoyed electricity in every house—in 15,000 altogether.

Sanitation. Cleaning services were delivered by the Municipal Cleansing Services which had 61 trucks, 36 sweeper-watering machines and horse-drawn carts with 90 horses. Snow was removed with a help of the unemployed who were paid in total for 90,000 working days in 1937.

2. Mobility

Every day to Warsaw 175 local trains and 74 intercity trains arrived. In one-year 1937, 4.5 ml. single tickets and 410,000 season tickets were sold. Trams network in 1937 was 244 km long which was not much shorter than in 2016 when it was 276 km long. Before WW II Warsaw had 29 daily lines and 9-night lines with 386 trams. Today there are 26 lines and 750 trams. In one-year 1937, 190 passengers used trams. The auto bus lines were also important means of transport. 65 buses on 14 lines carried 22 ml passengers a year. Cars were not so popular as there were only 3752 cars, 1251 tracks, 1563 motorcycles and 34,667 bicycles registered. They had to compete with 1500 carriages, 1258 horse-drawn wagons and 1453 other vehicles. Yet, there were gradually more taxi cars (2142) than horse-drawn taxis (1509). Warsaw had 141 petrol stations.

Communication. In 1937 there were 673 mailboxes and postmen in the city. That year citizens sent almost 250,000 letters. The telegrams proved also convenient as citizens received 1 ml telegrams and sent

581,000 messages. Telephone network was 1000 km long with 65,521 users and 81,000 telephones and only 318 public phone boxes.

3. Activities

Animals. In 1937, 9000 horses, 2100 cows, 1300 pigs, 500 goats and 100 sheep were reported in Warsaw. Whereas, every year 250,000 pigs, 160,000 calves, 83,000 horned cattle and 17,000 sheep were slaughtered.

Food and drinks. Varsovians ate 20 tons of butter and 11 tons of cream every day and drank 250,000 litres of milk. They consumed 21 tons of salt and almost 50,000 tons of sugar annually.

Smoking. In Warsaw in 1937 citizens bought legally 1,300,000 cigarettes, 2 million of cigars and over 400,000 kg of tobacco. They used 62 ml matches (probably also for other purposes).

4. Opportunities

Housing. In 1931 Warsaw had about 25,000 flats but 31 per cent of citizens lived in one-room flats (today's studio apartments). However, because of the huge housing crisis, these flats were occupied by 4 residents on average (Fig. 1.8).

Employment. Blue collar workers were the biggest group—52 per cent of all working citizens of Warsaw. White collar workers constituted 18 per cents and free lancers another 18 per cent. In Warsaw 65 per cent men and 34 per cent of women were working.

Weddings, births and deaths. There were 107,000 weddings organized annually, 16,000 children were born and 13–14,000 died including 1700 children below one year old.

5. Governance

Under the Act of 1923 on providing social security public institutions in Warsaw provided help to 20,000 homeless people in specially built shelters, which may serve as an example of modern pro-social governance (Kośka, 2017, Nov.11).

Fig. 1.8 The photo depicting the housing shortage in Warsaw after WW I (*Source* Wielka Historia. (n.d.). Przedwojenna Warszawa w liczbach. https://wielkahistoria.pl/przedwojenna-warszawa-w-liczbach-zaskakujace-fakty-i-statystki/)

1.6 Conclusion

Although the concept of smart city ecosystem is developed in the contemporary literature, its traces can be also identified in the pre-war cities. The apparent examples of smartness are the pro-social (human-centric) policy of shelters for unemployed and the well-developed network of public transport. The issue of affordability has also become crucial for contemporary growing and overpopulated cities, which structures have been shaped not only by prosperous entrepreneurs but also the influx of the underprivileged. The high-tech advances are to accelerate the sustainable growth of the cities which create complex and networked ecosystem enhancing the quality of life of dwellers and regulating its metabolism of natural resources. Availability of resources will determine the future city development which may take the defined-above, three completely different scenarios. One of them prophesizes the destruction of the operational foundations of the present world. This scenario triggers the further discussion on city infrastructure as a crucial supply network sustaining city life and a target of (Chapter 2) city warfare (Chapter 3 and 4) as a future form of wars.

References

Alberti, M. (2024). Cities of the Anthropocene: urban sustainability in an eco-evolutionary perspective. *Philosophical Transactions of the Royal Society B, 379*, 2022026. https://doi.org/10.1098/rstb.2022.0264

Alfsen, Ch., Ashley D., & T. Elmqvist (2011). The Urban Landscape as a Social-Ecological System for Governance of Ecosystem Services', in Jari Niemelä, and others (eds), *Urban Ecology: Patterns, Processes, and Applications*. (Oxford, 2011; online edn, Oxford Academic, 17 Dec. 2013). https://doi.org/10.1093/acprof:oso/9780199563562.003.0026.

Bettencourt, L.M.A., Lobo, J., Youn, H. (2013). *The hypothesis of urban scaling: formalization, implications and challenges*. arXiv:1301.5919v1 [physics.soc-ph], 24 Jan, https://doi.org/10.48550/arXiv.1301.5919

Borer, M. I. (2006). The Location of Culture: The Urban Culturalist Perspective. *City & Community, 5*(2), 173–197. https://doi.org/10.1111/j.1540-6040.2006.00168.x. Retrieved April 20, 2024, from https://www.researchgate.net/publication/229666310_The_Location_of_Culture_The_Urban_Culturalist_Perspective

Dear, M., & Flusty, S. (1998). Postmodern Urbanism. *Annals of the Association of American Geographers, 88*(1), 50–72. https://www.jstor.org/stable/2563976

Denton, N. A. (1994). Are African Americans Still Hypersegregated in 1990? In: R. Bullard, C. Lee, & J. E. Grigsby (Eds.), *Residential Apartheid: The American Legacy*. UCLA Center for Afro-American Studies, Los Angeles, pp. 49–81.

Dumont, G. (2013). L'urbanisation: Un processus global, une réalité locale. *Population & Avenir, 715*, 3–3. https://doi.org/10.3917/popav.715.0003

Farkas, K., Szabo, C. A., & Horvath, Z. (2009). *Planning of Wireless Community Networks*. Handbook of Research on Telecommunications Planning and Management for Business. https://doi.org/10.4018/978-1-60566-194-0.ch041

Fouberg, E. H., Murphy, A. B., & Blij, H. J. de .(1977). *Human geography: People, place, and culture*. (1st edn., Hoboken, N.J).

Graham, S., & Marvin, S. (1999). Planning Cyber Cities: Integrating Telecommunications into Urban Planning. *Town Planning Review*, 70, 89–114. https://doi.org/10.3828/tpr.70.1.w34454x3475g2858

Friedmann, J. (1986). The World City Hypothesis. *Development and Change, 17*(1), 69–83. https://doi.org/10.1111/j.1467-7660.1986.tb00231

Hall, P. (2001). Global city-regions in the 21st century, in Scott, A. J. (Ed.) *Global City-Regions: Trends, Theory, Policy*, pp. 59–77. Oxford University Press, New York, NY.

Harris, C. D., & Ullman, E. L. (1945). The Nature of Cities. *Annals of the American Academy of Political and Social Science, 242*, 7–17.

Hoyt, H. (1939). *The structure and growth of residential neighborhoods in American cities*. Washington, DC: Federal Housing Administration.

IIoT World. (2018). *The Smart City Ecosystem Framework—A Model for Planning Smart Cities*. January 26, 2018. Retrieved October 2, 2024, from https://www.iiot-world.com/smart-cities-buildings-infrastructure/smart-cities/the-smart-city-ecosystem-framework-a-model-for-planning-smart-cities/

Jaeger, B. (2009). *E-Government and E-Democracy in the Making*. Encyclopedia of Information Science and Technology, Second Edition. https://doi.org/10.4018/978-1-60566-026-4.ch208

Janoschka, M. (2002). The new model of the Latin American city: fragmentation and privatization. *EURE (Santiago) November, 28*(85), 11–29.

Jayaram, N. (2010). Revisiting the City: The Relevance of Urban Sociology Today. *Economic & Political Weekly, 45*(35), 49–57.

Josse, G. (2020). Planning and managing Africa's cities: what place for technology innovations? *Field Actions Science Reports* [Online], Special Issue 22. http://journals.openedition.org/factsreports/6317.

Kilcullen, D. (2015). *Out of the Mountains: The Coming Age of the Urban Guerrilla*. Oxford University Press.

Kośka M. (11.11.2017). *Bieda, przeludnienie i zacofanie wsi. Dwudziestolecie międzywojenne to nie tylko rozwój Gdyni i salonowe życie w Warszawie*; https://www.money.pl/gospodarka/wiadomosci/artykul/dwudziestolecie-miedzywojenne-cop-gdynia,172,0,2387628.html

Latham, A., & Layton, J. (2019). Social infrastructure and the public life of cities: Studying urban sociality and public spaces. *Geography Compass, 13*, e12444. https://doi.org/10.1111/gec3.12444

Lewis, P. F. (1983). The Galactic Metropolis. In R. Platt & G. Macinko (Eds.), *Beyond the Urban Fringe* (pp. 23–49). University of Minnesota Press.

Mead, W. R. (2024). *The Rise of the Cyber City*. Retrieved May 18, 2024, from https://www.hudson.org/politics-government/rise-cyber-city-walter-russell-mead

Moniruzzaman, M. Rahman, M.M. Tasfi, S.M. (2013). Exploring the Nature of Urban Structure of Urban Areas in Bangladesh. International Journal of Sciences. Volume 2, Issue Jan., 91–103.

Montgomery, J. (1998). Making a City Urbanity Vitality and Urban Design. *Journal of Urban Design, 3*(1), 93–116.

Montgomery, J. (2008). *The New Wealth of Cities City Dynamics and the Fifth Wave*. Routledge.

Neuman, M., & Hull, A. (2011). *The futures of the city regions*. Routlege.

Ortman, S. G., Lobo, J., & Smith, M. E. (2020). Cities: Complexity, theory and history. *PLoS One, 15*(12), e0243621. https://doi.org/10.1371/journal.pone.0243621. PMID: 33290411; PMCID: PMC7723246.

Park, R., & Burgess, E. W. (1925). *The City*. The University of Chicago Press.

Qadeer, M. A. (2024). Perspectives on cities of the global south. *Planning Theory*. https://doi.org/10.1177/14730952241276686

Siebel, W. (2018). Urbanität. In *ARL—Akademie für Raumforschung und Landesplanung (Hrsg.): Handwörterbuch der Stadt-und Raumentwicklung* (pp. 2755–2766). Hannover.

Sparks, K., Thakur, G., Pasarkar, A., & Urban, M. (2020). A global analysis of cities' geosocial temporal signatures for points of interest hours of operation. *International Journal of Geographical Information Science, 34*(4), 759–776. https://doi.org/10.1080/13658816.2019.1615069. https://wielkahistoria.pl/przedwojenna-warszawa-w-liczbach-zaskakujace-fakty-i-statystki/

UN (2018). The World's Cities in 2018.

UN (2022). World Cities Report 2022. Envisaging the Future of Cities. UNHabitat.

Wielka Historia. (n.d.). Przedwojenna Warszawa w liczbach. https://wielkahistoria.pl/przedwojenna-warszawa-w-liczbach-zaskakujace-fakty-i-statystki/

Wirth, L. (1938). Urbanism as a Way of Life. *American Journal of Sociology, 44*(1), 1–24. http://www.jstor.org/stable/2768119

Zeihan, P. (2022). *The End of the World Is Just the Beginning: Mapping the Collapse of Globalization*. HarperCollins.

Zhao, S. X., Guo, N. S., Li, C. L. K., & Smith, C. (2017). *Megacities, the World's Largest Cities Unleashed: Major Trends and Dynamics in Contemporary Global Urban Development* (Vol. 98., October 2017, pp. 257–289). World Development.

Zhu, D. (2009). Nature Inspired Parallel Computing. In *Handbook of Research on Artificial Immune Systems and Natural Computing: Applying Complex Adaptive Technologies* (pp. 496–519). https://doi.org/10.4018/978-1-60566-310-4.ch023. https://www.albert.io/blog/galactic-city-model-ap-human-geography-crash-course/

CHAPTER 2

Infrastructure—A Critical Element of City Ecosystem

Abstract The 21st city infrastructure is intertwined, complex and multifunctional and determines citizens' safety and quality of their lives. This chapter discusses critical infrastructure for citizens survival: water, energy, communication, housing and public infrastructure. It presents the example of smart and resilient infrastructural solutions, threats to this infrastructure, impact of its damage on citizens and management strategies in times of crisis (selected cases).

Keywords City infrastructure · Resilience · Threats to infrastructure

2.1 Infrastructure—A Main Security Asset

Urban assets are commonly referred to as urban infrastructure, which includes facilities and equipment for transportation, energy, water supply, sewage treatment and solid waste disposal or, in a more general sense, the built environment (Friedmann, 2005). They are intertwined, complex and multifunctional.

> The contemporary way to think about infrastructure is not as a single-dimensional thing, but in a multivalent, multidimensional way. Everything we design infrastructurally must have more than one facilitation. You see

S. Kowalkowski et al., *Cities in Crisis*,
https://doi.org/10.1007/978-3-032-08843-7_2

> that with public libraries now: It's not just a place that holds books; it's a place where people meet, learn things, attend events. It is, in some cases, an extension of the public realm. In the 20th century, we would think of infrastructural things as one-dimensional; in the 21st century, we have to think of them as multidimensional. (Cimino, 2017)

Twentieth-century infrastructure was about conquering the landscape. We built freeways, power plants and other necessities regardless of the natural environment; we conquered it. We would blow through mountains and we'd do whatever we needed to because we could, technologically, conquer nature. We're now recognizing—especially in the wake of the recent hurricanes Harvey, Irma and Maria—how that kind of mentality leaves you with an unresilient system. Twenty-first-century infrastructure should work with the natural landscape in ways that are not about conquering, but are about adaptation (Cimino, 2017).

Zeihan (2022) predicts highly developed cities will face consequences of depopulation, whereas the overpopulated cities of Asia and Africa may struggle with all the post global events and crises:

> If measures are not taken soon, some of these megacities will pose the most significant security threat in the coming decades. They will become havens for terrorists and criminal networks, as well as sources of major environmental depletion. They will serve as volatile natural laboratories where all the elements most harmful to international and human security are grown. (Liotta & Bilgin, 2014, p. 109)

Concurrent shocks, deeply interconnected risks and eroding resilience are giving rise to the risk of polycrises—where disparate crises interact such that the overall impact far exceeds the sum of each part. These potential polycrises resulting from disintegrated geopolitical cooperation will be of complex environmental, geopolitical and socioeconomic nature and they will relate to imbalance of the supply and demand for natural resources (WEF, 2023, p. 9).

Method: The-above-pictured security environment in which supply of resources play the decisive role, the infrastructure has become the critical structures upon which cities may develop, respond to ongoing crises and survive. As never before they rely on interconnected, complex and interdependent networks and systems: energy, water, communication, transactional infrastructure, civil security and law enforcement agencies, road traffic, public interest networks and services offered by public and private operators (CORDIS). We adopt the analytical pattern (infrastructure malfunctions—impact of its damage—countermeasures) in our research. Referring to the reports, legal documents, literature, cases of given crisis situations and implemented response solutions we address the following problems: how functioning of city infrastructure is connected with physical and social environment, how both natural and human driven disasters/crises may disturb its functioning and what impact this damage has on city dwellers and finally, what exemplary practical resilience building solutions have been adopted to prepare for or mitigate the crises.

2.2 Water Infrastructure

Urbanization significantly impacts water resources in many ways. Impervious surfaces of man-made constructions (e.g.: roads, buildings, parking lots) and over-extraction of groundwater cause flooding whereas industrial and agricultural practices, waste disposal and sewage cause accumulation of pollutants and consequently degradation of water quality (Bilgiç & Baba, 2023). Mismanagement coupled by climate change (intense floods and prolonged droughts) and an increase in water demand for overpopulated megacities will reduce opportunities to cooperate over shared water resources (Zawahri & Weinthal, 2019). It is projected that by 2050, 2.065 (1.693–2.373) billion people will suffer from water scarcity; 840 (476–905) million people will face perennial water scarcity and 1.225 (0.902–1.647) billion will face seasonal water scarcity. India's growth in water-scarce regions is to be much higher than other countries, increasing from 222 million people to 550 (376–644) million people in 2050 and accounting for 26.7% (19.2–31.2%) of the world's urban population facing water scarcity. The megacities facing water scarcity include: Cairo, Dhaka, Jakarta, Lima, Manila, Mumbai, New York, Sao Paulo, Shanghai and Tianjin . It is projected that 44 counties will have faced the high or extremely high level of water stress by 2040 (Fig. 2.1).

The cascading effect of the climate change starts with the disruptions in everyday life in cities through extreme events like floods and landslides and slow-moving stressors like water scarcity and population displacement. These effects spread through global and interconnected supply chains and infrastructure networks reaching distant locations in different parts of the world and generating or amplifying other unpredictable hazards (from complex adaptive systems theory and non-linear

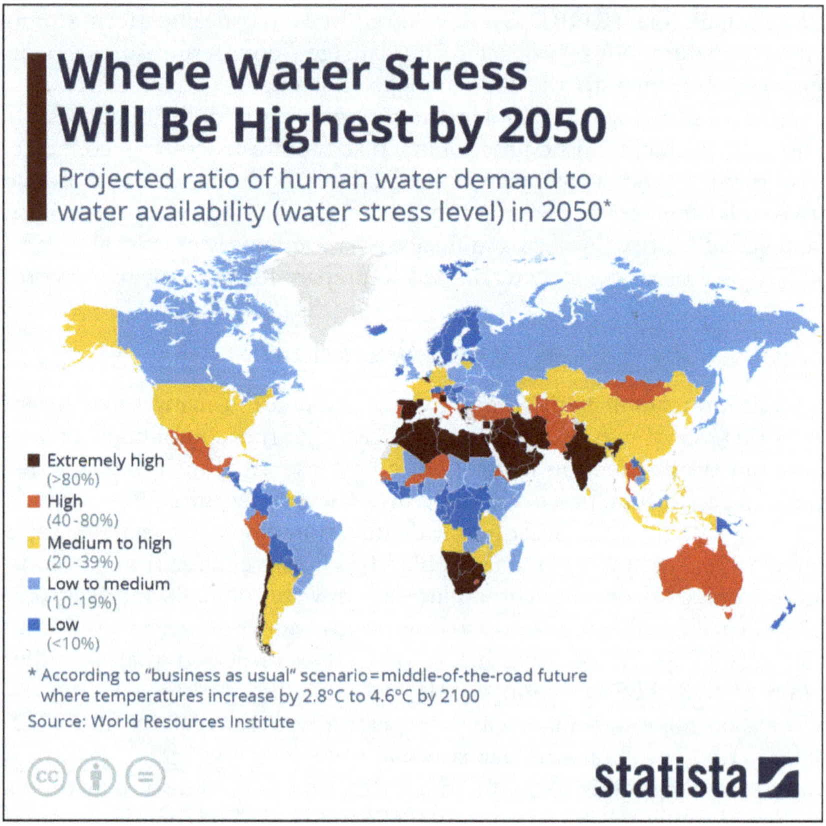

Fig. 2.1 Water stress by 2050 (*Source* https://www.statista.com/chart/26140/water-stress-projections-global/)

feedback dynamics in Westman et al., 2022; Thailand 2011 floods case in Bowen et al., 2022).

Moreover, drought and water shortages accompanied with regional violence and political turmoil may contribute to internal instabilities. Water is one of the drivers of the decades-old conflict between Palestine and Israel. Saudi Arabia's government said its people will depend entirely on grain imports by 2016, due to fear of water-resource depletion. Water problems will put key North African and Middle Eastern countries at greater risk of instability and state failure and distract them from foreign policy engagements. Water scarcity and chronic mismanagement forced 1.5 million people, primarily farmers and herders, to lose their livelihoods and leave their land, move to urban areas and magnify Syria's general unrest and civil war in Syria 2011 (World Resources Institute, 2015).

Planning, budgets and infrastructure in overpopulated megacities are failing to serve most urban residents in the developing world with water and sanitation, contributing to poor health conditions and heavy pollution loads in wastewater. Unsanitary conditions create a constant threat of a disease outbreak, such as cholera, that can devastate poor communities and spread through the city and beyond. Moreover, the poorest neighbourhoods are often in areas more exposed to natural hazards such as flooding, sea level rise, forest fires, landslides, volcanic eruptions and tsunamis. A lack of robust water and sanitation systems and hygiene facilities weaken even more the community's ability to stay healthy during environmental shocks (UN Water).

Urban water scarcity has been addressed in many ways: developing technologically advanced engineering and infrastructure; building reservoirs to store water during periods of excess availability and supply water to cities avoiding water shortages during dry periods; desalinating plants to solve water deficit problems for coastal cities; or inter-basin water transferring. However, developing water infrastructure requires considerable investments and is conditioned by natural characteristics, topography and geographic location and may have significant environmental impacts (He et al., 2021).

Japan, for instance, affected by the climate and natural environment, water raises challenges elaborates the highly advanced solutions to ensure a sustainable water cycle and robust water infrastructure. Hitachi offers a comprehensive set of variety of technologies, systems and services to address challenges of water management (operations, planning, maintenance, monitoring, water treatment and equipment): conservation of water resources, flood control, water use, domestic and industrial water supplies, sewage and industrial wastewater treatment and water production and recycling (Fig. 2.2).

Following the concept of Society 5.0—"a prosperous and human-centric "super-smart society" that takes maximum advantage of information and communication technology (ICT)" (Tachi et al., 2019, p. 59). Water environment solutions combine operational technology (OT), information technology (IT) and products to provide robust water

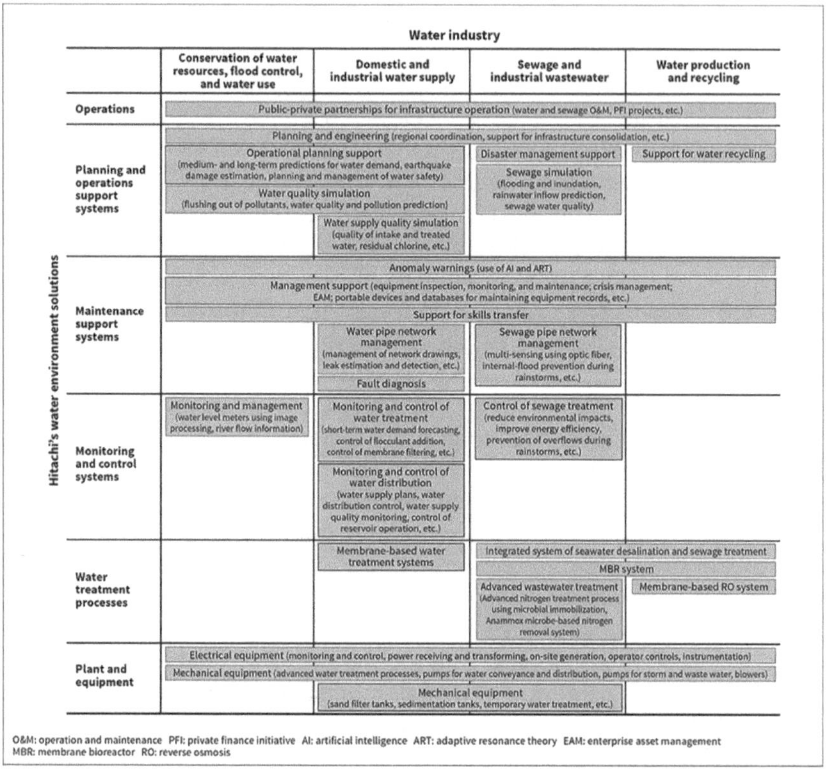

Fig. 2.2 Sustainable water cycle (*Source* T. Tachi et al. [2019]. Overview and Examples of Water Infrastructure Solutions, Hitachi Review Vol. 68, No. 5. https://www.hitachihyoron.com/rev/archive/2019/r2019_05/05a01/index.html)

infrastructure, improve people's QoL and overcome societal challenges. The list of innovations includes:

- On-site, multipurpose water quality gauges for making various real-time measurements across multiple locations of tap water quality, including the concentration of residual chlorine;

- The energy-efficient water treatment technologies help reduce the load on the environment, including membrane separation, separation and recovery of valuable metal and toxic materials and the decomposition of substances that are difficult to break down;
- The IoT is used to collect various data on the operations of water and sewerage utilities and save it in the cloud. This service supports skills transfer and visualization, labour saving and efficiency improvement in operation and maintenance;
- Multi Sensor boxes installed on the optic fibre network that runs through sewage pipes, facilitates water level, water quality, anti-flooding measurements and pipe network management remotely over a wide area. The external power supply has been replaced by low-power sensors powered by the optical energy from the communication signals (Tachi et al., 2019).

Shrinking water resources and impaired water supply infrastructure may lead to societal and economic and energy problems. Water resources are generally discussed in context of natural disaster; however, the focus should be placed also on dependent energy production, fuel consumption and prices (Gazeta Prawna, 2021; World Bank, 2014). Central Europe, Poland also suffers from seasonal water scarcity. In Świnoujście, a tourist resort during summer, when water consumption increases, the residents suffer from shortages of tap water due to low levels of groundwater (Wysota, 2023). Temporary water shortages may affect also other parts of Poland, especially central provinces like łódzkie and greater Poland and different industries: mining, metallurgic, chemical, paper making, food, oil-refining.

Impaired water infrastructure is a concurrent issue. 67% of infrastructure is 40–60 years old (NIK, 2018), which consequently accounts for numerous but tot long-lasting system failures, flooding of roads during heavy rainfalls and 30% water loss out of water supply to households. 25% of citizens still do not have access to a sanitation system.

Development of water infrastructure is supported by several investments. In Ząbki, Poland the modern water treatment plant will be open in 2024 to provide drinking water to 24,000 citizens; in Elbląg a new sewage pumping station maintains 73% of town sewage.

Moreover, an intelligent remediation system for neutralization of harmful contaminants in water, based on carbon electrodes is being developed by the Polish-Norwegian consortium, including experts from

Gdańsk University of Technology. An energy-efficient system with AI support can be used to treat larger mass of sewage, e.g. in treatment plants. The system will be designed to enable oxidation of pollutants and a shorter waste disposal chain (Gdańsk Tech). The investment in inland waterway transport infrastructure in Poland increased by 25.3 million euros (+64.61%) since 2007, reaching its highest value in the observed period. In 2022, the length of the water supply network reached 337.7 thousand kilometres, an increase of 36.7 thousand kilometres compared to 2016. Inland waterway infrastructure investment is expanding, updating and upgrading existing networks including locks, dams and navigation channels (Statista).

2.2.1 *Water Weaponization*

The first reported case of the weaponization of water was during World War I when Belgium city Niewpoort successfully used gates on the Yser River to hinder the advancement of the German army (Paraszczuk). Nowadays water resources in war can be instrumentalized in several ways: for domination and legitimacy, as military tools, military targets and cooperation. Marwa Daoudy (2020) discusses the Syrian conflict as an example.

When water resources and infrastructures are weapons of domination and legitimacy water, for value as a source of livelihood, becomes a symbol of identity and a mechanism of power consolidation. Non-state actors use the same strategies to legitimize their control over this infrastructure (e.g. ISIS after conquering Raqqa in January 2014 threatened population with punishment for electricity theft and water tax evasion). When water is a military target, the water systems are destroyed causing excessive damage to populations (e.g. On 23 December 2016 Syrian government destroyed Fijeh spring in besieged eastern Ghouta—'purposely'). Water infrastructure is weaponized as a military tool when a state or substate actor use water assets to terrorize populations, obtain concessions from opponents or support tactical goals of military actions (e.g. In April 2014 ISIS almost stopped the flow of water to four predominantly Shi'a cities and displaced 40,000 people in Abu Ghraib and its farmlands). Water cooperation in times of conflict can both cause harm to civilian populations and facilitate military goals (Over 2013–2017 Syrian government cooperated with ISIS in production and delivery to Damascus of hydroelectricity from Tabqa dam; employees were paid by central government).

Another classification of water weaponization relates to political, tactical and psychological settings. The political dimension entails exerting pressure on the opposing leader or on local populations to surrender or suffer from the lack of water. The tactical approach means in other words "water as a weapon" when water and water infrastructure is a tool to gain military advantage for example, by flooding an area to block the route for invading opponents (e.g. Chinese in 1938 during the Second Sino-Japanese War). The potential lack of water has also a strong psychological influence on local communities who would suffer extensively from multiple consequences as health disorders, famine, diseases.

The Islamic State models of instrumentalizing water exemplify the practices of the weaponization of water in the region (https://waterandconflict.web.unc.edu/weaponization-of-water/). A major water pump in the city of Maskanah, Syria supplies Aleppo with water. In December 2016 ISIS took control of the pump and turned it off, limiting the supply of water to Aleppo. In retaliation, the unidentified (though anti-ISIS) aircraft bombed the area controlled by ISIS including Maskanah to regain control of this area. The water supply returned to people but it is unclear who currently controls the city.

Libya also provides multiple examples how the infrastructure is targeted since there are different actors who have different interests not only in destroying infrastructure, but in capturing and controlling it. Water is used to extract demands or concessions from warring groups, militias and warlords. These tactics involve small-scale looting, sabotage, threats to personnel and obstructing access to repairing infrastructure, rather than targeted destruction.

The case seems to be different with Libya's main water utility, the Great Man-Made River Project (GMR) that supplied irrigation and drinking water to cities and farms in the more heavily populated north. There is, as local authorities claim, understanding that the infrastructure need to function to provide water to the population throughout the country and the GMR cannot be a tool for war. However, this highly centralized and networked system is also highly vulnerable in times of a protracted conflict when infrastructure becomes difficult or impossible to maintain: power outages can cause disruptions, spare parts are almost unavailable and system deteriorates; when the staff is taken hostage the operations are disrupted, parts of the system are looted (https://waterandconflict.web.unc.edu/weaponization-of-water/).

2.3 Energy Infrastructure

Cities, as part of global economy largely depend on fossil fuels. Consuming about 78% of the world's energy they account for more than 60% of global greenhouse gas emissions. Transition to net zero is determined by cities capacity to use alternative fuels to power their economies and communities.

> What's needed now is to accelerate a series of holistic shifts. Shifts are needed in the fuels that are used to power these systems (away from coal, oil and natural gas towards renewable electricity and green gases, such as hydrogen); in the technologies that make these systems run; and in the people who can install and manage them. (PwC)

Technologies include comprehensive solutions: more and more resilient platforms, microgrids and renewable energy systems, photovoltaic (PV) elements integrated into building design, heat pumps, highly efficient solar panels and wind power which become competitive with fossil fuels. Improved fuel cell and battery technologies foster the adoption of EVs and solar power. The 5G together with the cost-effective internet of things, smart metering and the monitoring of urban energy consumption will drive the expansion of smart energy grids market. Thus, by diversifying sources of energy, cities will be able to enhance their resilience. The implementation of diversification policies requires cooperation in public–private partnership (PPP) formula (Table 2.1).

To create value the stakeholders, need to interact in an expanding and interdependent ecosystem. Regulators and policymakers create the favourable environment for financing institutions to make long-term investments by promoting cleaner and more resilient energy generation, transmission and distribution. Energy service providers collaborate with R&D institutions to meet targets and develop innovations and turn to financing institutions for the necessary financial means. Financing institutions, in turn, can provide funding for clean-energy and energy-efficiency projects. Multilateral organizations support cross-border and cross-industry planning. End users—conscious key players—create demand and respond to incentives for new types of products and services.

Table 2.1 Energy transition—a cross-sectoral formula for diversification of energy supply

Regulators/policymakers			
Energy supply			*Energy demand*
Generation	*Transmission*	*Distribution*	*End usage*
Energy service providers Multilateral organisations/associations Financing institutions Research and development institutions			Consumers

Source Based on PwC Global Report: The Energy Transition. Building sustainable cities: How urban infrastructure can address energy challenges and shocks. https://www.pwc.com/gx/en/issues/esg/the-energy-transition/sustainable-cities-tackling-climate-change-through-urban-energy-transition.html (2 June 2024)

Taking charge of your energy demand can improve productivity and profit margins, reduce exposure to volatility and political events. Adaptable solution addresses the needs of cities in time of peace, crisis, shock or prolonged compound crisis. However, the time of war create extreme environment in which any system may become a target and fail and usually does.

2.4 Communication Infrastructure

Communication networks—hardware and software systems that facilitate communication of any form (video, text, data, or any other digital objects) from a local area network (LAN) to wide area network (WAN)—constitute a fabric underpinning social order and determining social dynamics. ICT infrastructure integrates interactions of smart city components (applications/services and the city infrastructures, such as electricity grid, roads, parking) either in form of human-to-human, human-to-machine, or machine-to-machine communication

> The core of the smart city information system is the pooled database that stores and correlates the data reported by the networks [...] The best solution for a municipality is a combination of mobile, fibre, Wi-Fi and Lora. Each technology meets a specific need, in terms of coverage, reliability and efficiency. (Stéphan Guidarini for Agility Effect, 2019)

Smart cities characterize with a large number of population and unevenly distributed infrastructures, which results in a large number of applications ends-points and variety of technologies applied.

The true enabler of smart city sustainability (and thus resilience and relative security) is Internet of Things (IoT). IoT is a network of physical objects that are embedded with electronics, software, sensors and connectivity. These objects can collect and exchange data with other objects and systems. IoT in smart cities can be used to monitor traffic congestion, air quality or noise pollution levels in the city. It can also be used to monitor the water levels in reservoirs or detect when a tree falls down on the street. The are other exemplary applications of IoT in cities (Haidine et al., 2016):

- Smart Lighting systems use sensors and automation to adjust the brightness of lights based on ambient light and traffic levels. This results in energy savings and improved safety for pedestrians and motorists.
- Intelligent Traffic Management uses real-time monitoring and control to optimize traffic flow, reducing congestion and improving safety for drivers and pedestrians.
- Smart Waste Management systems use sensors and real-time monitoring to schedule waste collection, reducing costs and improving the efficiency of waste management operations.
- Environmental Monitoring systems use sensors to continuously monitor air quality, temperature and other environmental parameters, providing information to help maintain a healthy environment.
- Smart Grid and Energy Management systems optimize energy distribution, monitor energy consumption and implement demand-side management to reduce energy waste and costs.
- Intelligent Parking Management systems use real-time monitoring and dynamic pricing to manage parking demand, reducing congestion and improving accessibility.
- Public Safety and Emergency Response systems integrate surveillance cameras, smart sensors and emergency communication to improve response time and efficiency during emergencies.
- Smart Buildings and Home Automation systems provide automated control of lighting, heating, cooling and other building systems to improve comfort, security and energy efficiency.

- Health Monitoring and Care Services use continuous monitoring of vital signs, fall detection and medication management to provide timely and effective care for elderly or disabled individuals.
- Smart Retail and Marketing systems use customer behaviour and preferences to offer personalized marketing and recommendations, as well as real-time inventory management to improve efficiency and customer satisfaction.

IoT connects "all real-world assets, equipped with a variable intelligence enabling information exchange, interaction with people, enterprises business process support and creativity" (Haidine et al., 2016, Ch. 5.1).

ICT platforms generally designed and exploited for variety of distributed uses (in an environment in which various components are deployed on several computers (including virtual machines) within a single network. These machines (which can be PCs, servers, or containers) distribute tasks among themselves and coordinate their actions for better efficiency/performance. During COVID-19 pandemic ICT were repurposed to maintain social connections, provide distributed services, continue to meet business needs and for virtual education and first of all to maintain social order. Using ICT included in particular: involving new entities in humanitarian collaboration networks (e.g. volunteers, nonprofit organizations, non-state actors); facilitating new forms of disaster relief actions (e.g. digital humanitarians, online self-support groups); and enabling diverse civic engagement (e.g. digital archives of deleted posts to counter censorship and disinformation campaigns). The COVID-19 pandemic is perceived as an information crisis (Xie et al., 2020) and demonstrates the limitations of ICT, especially social media: insufficient information for decision making, unreliable information for sound public debate, inaccessible information to meet people's daily information needs and misinformation, disinformation and fake news challenging citizen's health information literacy (Yang et al., 2020).

This case reminds also about seemingly obvious fact that smart city represents a purely human construct. An artificially designed material world incorporated in the natural landscape provides environment well-suited for the development of its political, cultural and economic activities (Marchal & Stebé, 2014).

> By essence, cities are communication hubs and have always brought people closer and made exchanges, interactions, transactions and of course separations possible. […] cities cannot do without communication and indeed find their very existence in migratory fluxes, diasporas and networks of exchanges firstly on the basic level of obtaining supplies and essential objects (Marchal & Stebé, 2014. p.2).

ICT with its endless lengths of cables, numerous channels of *wifi* wavelengths and infinite electronic networks constitute the communication structure of the smart digital city—"cyberterritory" which is a mirror reflection of the reality where the citizen operates in production, consumption, values, information exchange and government (Fig. 2.3). ICTs develop communication between citizens (social connectivity), support public life and democratic governance, trust-based relations between the elites and citizens and between public sector institutions and the private sector, consequently strengthening social cohesion. For that purpose, it is crucial to pay more attention to collective actions which involves conflicts, disputes, consensus or ambivalence.

Activities on a cyberterritory have their impact in the real world causing both worlds gradually merge. This happens when millions of individuals are sharing the same moment when entertaining during the Olympic Games, or dramatic—the 9/11 terrorist attack, or when "over 10,000 people send text or e-mail invitations to a chic 'dressed in white' dinner in a top Parisian location revealed just a few hours before the banquet".

The Oder Basin contamination in 2022 is another exemplification of how the media on one hand reflect and, on another shape, public emotions and behaviours. It may be clearly demonstrated by the fact that the disaster has different names in Poland and in Germany. In Poland it's mainly called "Katastrofa Ekologiczna" Ecological/Environmental Disaster, whereas on German internet its usually called "Fischsterben" (Fish kill). However, the popularity of each term was changing with time (Fig. 2.4)

The research on the Oder River crisis in media (Jarynowski, 2022) shows that SMS alarm on 12.08 was effective crisis communication tool in Poland; even refugees and migrant community became involved as an increase in reactions in Ukrainian language was reported. Another research on online public discourse on the Oder River crisis (Grelowska et al., 2024) tracks how emotions and discourse are evolving throughout

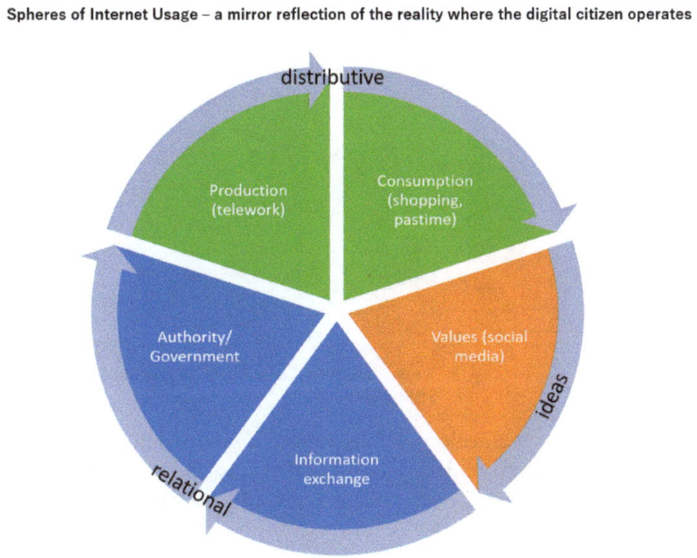

Fig. 2.3 Spheres of Internet Usage (*Source* Own elaboration based on: Moody and White [2003] and Szpunar [2008])

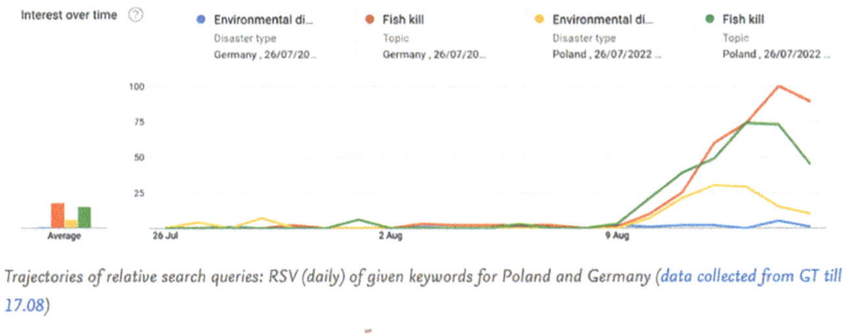

Fig. 2.4 Trajectory of terms used (*Source* A. Jarynowski, Oder River disaster in media, Infodemics. Freie Universitat, Berlin. https://blogs.fu-berlin.de/infodemic-corona/2022/08/19/oder-river-disaster-in-media/)

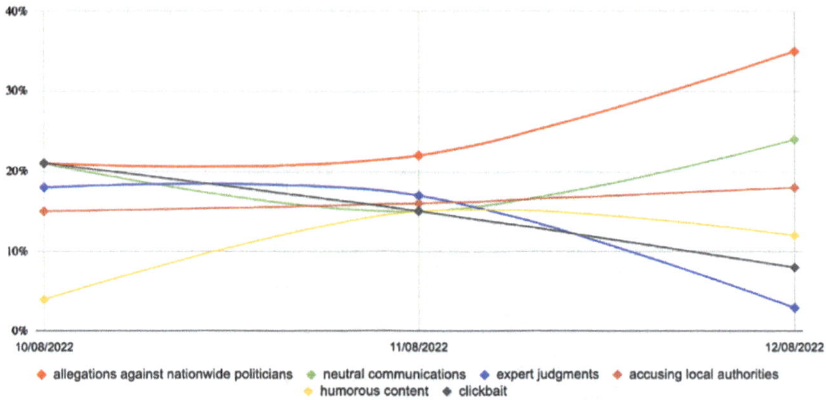

Fig. 2.5 Trajectory of emotions in discourse (*Source* Grelowska, M., Hibner, M., Deneka, W., Jarynowski, A. [2024]. Information Flow During Culmination of Online Public Discourse Based on The Oder River Environmental Disaster)

the crisis (Fig. 2.5). The amount of 54% of the analysed data carries a negative sentiment, attracting more engagement on social media, whereas 42% is neutral and the remaining 4% conveys a positive sentiment.

Posts with humour with a scent of irony communicate absurdities that could hinder assistance to communities or even create health and safety risks. The public interest evolves from a search for factual information, the creation of humourous content, threat evaluation to moving away from expert advice towards independent criticism of that.

The Oder River crisis revealed also how important is an effective DRM information system. The delays in communication unleashed the disinformation and fake news that bred unrest in local communities and hindered cooperation with NGOs and undermined the relations between public and non-profit organizations (Kowalkowski et al., 2024).

ICTs, in spite of weaknesses, they are designed and develop to effectively support the communication and cooperation of all entities that respond to crisis situations.

2.4.1 CMS JASMINE—Multilevel Crisis Management System

Multilevel Crisis Management System JASMINE (CMS JASMINE) is dedicated to NATO, EU and national crisis management structures. The system provides the comprehensive ICT support of public administration, security services (the Police, Fire Brigade, Border Guard, Emergency Medical Services, Crisis Management Centers) and Armed Forces in crisis operations, rescue and preventive actions. The solution enhances the management, planning, command and control as well as monitoring of political, military and non-military threats.

Its main functionalities are:

- planning, command support, coordination and control of operations (calculation, verification and simulation of courses of action, logistic plans, etc.);
- gathering, processing, aggregation and distribution of information about operations:
  - incidents (e.g. disturbance of security and public order),
  - events (e.g. pollution of water intake and forest fire),
  - threats (e.g.: epidemiological, terrorist, migration and constructional),
  - terrain (e.g.: flood areas, embankment status and chemical contamination),
  - population (e.g. location and number of people staying in buildings or regions of operation to provide assistance),
  - resources (e.g.: human, objects, units, protective equipment);
- current and automatic reports about last actions and available resources;
- distribution of information (including alerts, messages, reports, orders) to selected groups of recipients (troops, rescue teams, people) in any localization;
- cooperation with other systems of governmental organizations (Government Security Centre, Institute of Meteorology and Water Management, National Fire Brigade, the Police, National Emergency Medical Services, Border Guards) via application of centralized WWW portal—JASMINE Web Portal through sharing documents and information with CMS JASMINE;
- high multi-level security of information and data processing;
- presenting and monitoring of current crisis situations on digital maps based on GPS information, data entered by soldiers/officers/rescuers, as well as video images and photos from the actions;

- current video streaming from areas of: border lock, quarantine, search/rescue operations, fires, flood, etc.;

Examples of imaging non-military threats in Crisis Management System JASMINE are presented in the scheme below (Fig. 2.6).

The *Jaśmin* Crisis Management System was used, for example, to position medical volunteers during the UEFA U-21 tournament in Poland in 2017, to support the accommodation process (e.g. in hotels, guesthouses and other accommodation facilities) of refugees from Ukraine by collecting and monitoring information about vacancies. The *Jaśmin* is discusses in plans of building offshore wind farms (MFW) as it can offer comprehensive, automated management and protection against multiple, potential threats (terrorist attacks, protests, vandalism and force of nature) (ZBiAM, 2022).

Fig. 2.6 Examples of imaging non-military threats in Crisis Management System JASMINE (*Source* TELDAT. https://www.teldat.com.pl/en/offer/products/systems/330-cms-jasmine.html)

2.5 Shelters

The UNHCR Shelter and Settlement Global Strategy focuses on two mutually reinforcing objectives that aim to improve the living conditions of refugees and other People of Concern (PoCs) affected by conflicts and disasters aggravated by the compound effects of migration, food insecurity, urban vulnerability, climate change, environmental degradation and who are seeking protection and assistance in rural and urban settings. These objectives are:

> To enable refugees to access and live in dignity in secure settlements that prevent and reduce underlying causes of socio-economic vulnerabilities and allow for improved quality of life.
> To enable refugees to access shelter solutions which provide privacy, security, protection from the elements, emotional support and a space to live and store belongings in a dignified manner. (pp. 6, 10)

The five principles that guide realization of UNHCR shelter and settlement strategic goals are:

- Human rights—the right to adequate housing Article 25 (1) of the Universal Declaration of Human Rights;
- Age, Gender and Diversity—respecting diversity of the communities, their gender equality and their access to rights despite the age and the cultural, religious and educational background and particularly women, children, the elderly, persons with disabilities and minority groups;
- Participation—refugees and others of concern should be empowered at all stages of program/project design. They should be also engaged in the design, construction and maintenance of any shelter, settlement and CRI support.
- Sustainability—programs should be developed with sustainability solutions: appropriate technology, capacity building of both people of concern and local communities and use of local skills, materials, techniques and knowledge.
- Environmental Considerations—shelter and settlement interventions need to be planned and implemented with respect to regulations and in the manner to mitigate, to the extent possible, the impact on the

natural environment and to prevent hazard risks: landslides, floods, earthquakes.

In 2021 UNHCR in collaboration with government and other partners offered settlement and shelters to over 51,000 Persons of Concern (PoCs) in Dzaleka, Malawi refugee camp. The camp covers a space of 209 hectares and was initially designed to accommodate 10,000–12,000 individuals; however, it is now home to more than 51,000, which of course results into varying constraints (UN Partners Portal, 2024). This is also the similar case in Bangladesh where in 2018–2019 UNHCR effectively delivered infrastructure ensuring 100% coverage of adequate "temporary" shelters. Generally, the minimum standard for internal shelter living space is respected, yet, the external space is below acceptable levels, which generates increased protection and health risks. Moreover, the difficult topography especially during the monsoon season limits access to services. In 2020, the UNHCR focused on three areas: improved shelters, maintenance of temporary shelters and construction and maintenance of infrastructure. Shelter fundraising and advocacy gave links between shelter, protection benefits and health outcomes (UNHCR Global Focus).

The protection of civilians in the areas affected by war heavily depends on effective critical infrastructure shelters. Critical infrastructure is generally referred to government or military command centres, but it also includes municipal utility suppliers, public services health-care centres, communication towers, data centres and shelters which have to operate in crisis situation. Critical infrastructure shelters differ from regular civil shelters. They are usually fully operational in peacetime and should be transformed into a fully operational protected facility rapidly if necessary. These shelters also have different protection requirements as they are more likely to be targeted and the inhabitants are not there to rest but to work under pressure and therefore, they require more fresh air and more air changes per hour (Temet).

> A home bunker or shelter is a secure and fortified space designed to protect individuals and their families from various threats, such as natural disasters, terrorist attacks, pandemics and even war. These shelters can range from small underground bunkers to large-scale fortified rooms within a home or building. (European Shelters)

Ukraine in case of crisis situation has 5704 bomb shelters and 15,393 fallout shelters.

A fallout shelter is a category of shelter specially constructed of steel and located underground to protect people from a nuclear blast. The deeper the shelter is built, the safer inhabitants are. However, no area within long distance will be completely free from radiation. (US SafeRoom). The capital city of Kiev has 514 shelters including also that still under construction and partially completed and prepared as well as 4400 dual use structures: cellars, underground car parks, metro, underpasses. The interactive map of shelters in Kiev is available at the governmental sites (Bogarov, 2024, April 26).

Odessa has 353 shelters and about 800 dual-use structures, Kharkiv has only 300 bomb shelters and 4600 dual-use structures, Zaporizhzhia has 273 bomb shelters and about 3000 dual-use structures and Rivne has 197 shelters. In other region of Ukraine there are fewer shelters, for example Lviv has only 70 shelters and 6000 dual-use structures.

In time of peace shelters may be exploited as public facilities, for example: sanitary facilities, cultural sites, manufacturing workshops with the maximum D fire category, tunnels, passages, car parks, warehouses for fireproof materials, eating houses, sports and entertainment facilities.

Requirements and standards for bomb shelters in Ukraine may serve as a good practice (State Emergency Service of Ukraine No 03-1870/162-2, 14th June 2022; Wysocki). Shelters should ensure the protection of the population from the following hazards:

the effects of an air shock wave, conventional means of destruction, negative impact from other buildings, structures and from external ionizing radiation. In order to ensure these kinds of protection, shelters should meet the following requirements:

- Bomb shelter should be located in a cellar or cellar in the building. With the consent of the state authorities the first floor of a building may also be adapted.
- Deep bomb shelter and shelters with elevation may be freestanding structures,
- Shelters may be organized in schools, libraries, hangars and greenhouses,
- one 2 m^2 sanitary point should be set for every 500 people in a shelter and one additional 9 m^2 medical point—for 900 people.

- Shelter may house number of people calculated by summing up the surface necessary for sleeping and sitting: 0.6 m^2 of the area per person. If it may be furnished with bunk beds, the area can be reduced to 0.5 m^2 per person, with a triple bunk bad—up to 0.4 m^2 per person. The shelter organized in the already existing structure can house 5–50 people and should have at least one emergency exit, whereas 50 people or more can be located in a freestanding building with at least two emergency exits.

Shelters are not undergoing negative impact of ground, surface, technological or sewage waters. They are provided with electricity, artificial lightning, water supply and sewage systems. In the absence of water supply and sewage facilities, there should be the separate room for the portable sewage tanks and have forced or natural ventilation. Water and sewage systems cannot run through the premises intended for the population to be sheltered.

Shelters are easily accessible for the disabled.

Openings are closed by reinforced door made of non-combustible materials. Walls should be made of 2–2.5 thick brick or solid reinforced concrete structures (blocks, panels) 56 cm thick. Shelter should be also protected/covered with a layer of soli 67–78 cm thick. If the shelter structure is not thick enough it should be strengthened with additional layer of bags of soil, screens of reinforced structures or the mix of all (Fig. 2.7).

For comparison, the fragments of Fort Worth, Texas regulations on shelters are recalled below (FORT WORTH, TEXAS CODE OF ORDINANCES; 2024 S-16 Supplement). Each civil defence fallout shelter shall be subject to the following standards and regulations.

Maintenance and exploitation.

Civil defence fallout shelters shall:

- provide a flow of three cubic feet of fresh filtered air per minute for each person occupying the civil defence fallout shelter
- be maintained in a safe and sanitary manner and shall not be used for the storage of material other than those considered desirable for use as a civil defence fallout shelter.
- have a minimum amount of floor space as follows (Table 2.2).
- Storage of fresh water shall be provided on a basis of one-half gallon per person per day.

Fig. 2.7 Exemplary shelter in Poland (*Source* Szlak Podziemi. http://szlakpodziemi.pl/)

Table 2.2 A minimum amount of floor space (2024 S-16 Supplement)

Rated capacity in persons	*Sq. Ft. living area per person (based on stacked bunks)*	*Sq. ft. storage space per person*	*Minimum sq. ft. required*
1	24 (7.3 m^2)	2	26 (7.92 m^2)
2	12 (3.65 m^2)	2	28 (8.53 m^2)
3	10 (3 m^2)	2	36 (10.97 m^2)
4 (or more)	10 (3 m^2)	2	–

- Sanitary facilities shall not be connected directly to the city sanitary sewer system or to any septic tank sewer system. Each civil defence fallout shelter shall have a chemical toilet facility or an adequate supply of plastic bags or closed containers to serve the number of persons to be housed.

Construction. Civil defence fallout shelters shall.

- be designed to withstand displacement which might be caused by buoyancy produced by the surrounding fluid materials;
- may be constructed of masonry, reinforced concrete, steel, plastic or other approved materials.
- The walls and floor slabs shall be waterproofed using either a plastic membrane, mopped on asphalt or integral mixed water-proofing to resist the pressures to which they may be subjected;

In Poland the Supreme Audit Office (NIK) reported that less than 4% of Poles would find refuge in shelters. Another survey informed that the available bomb shelters could only accommodate 300,000, other secure spaces, such as metro stations and tunnels, could hold up to 50 million people. Whereas the report of State Fire Service on building inventory reads that 234,735 buildings across country were examined. These buildings are classified as protective structures i.e. shelters, places of hiding and places of emergency shelter. The biggest number of buildings were estimated in Mazovian province (29,700) and Silesia (25,900). Generally, in Poland the inventory list includes:

- 224,113 places of emergency shelter,
- 10,622 protective structures, including:
  - 903—shelters,
  - 8719—places of hiding.

Capacity of these buildings exceeds the population of Poland. Places of emergency shelter, shelters and places of hiding can accommodate over 49 m people (KGPSP).

Moreover, to address the needs of citizens Warsaw Major, for example, announced the 117-million-zloty (€27 million) program Warsaw Protects to improve the capital's emergency preparedness, including a better system of shelters. At present in Warsaw there are seven million square metres of underground car parks, the metro and all sorts of underground structures to serve as shelters. The Major "wants there to be an independent water and energy link in every school" and a reinforced shelter in all new buildings, storage facilities with equipment that can be used in emergencies such as floods, as well as improved systems for warning residents. The annual education week and safety training for citizens will be organized. An online service (available at schrony.straz.gov.pl) that helps people find their nearest place of refuge in an emergency has been already launched.

War in Gaza, 2024 demonstrates the cities infrastructure vulnerabilities and importance explicitly. According to the Government Media Office as of 7 January over 69,000 housing units were totally destroyed and over 290,000 housing units damaged, over 80% of buildings in the two northern governorates were uninhabitable. The destruction of central and southern Gaza was so large that any possibility of return would first require explosive ordinance clearance, debris removal and infrastructure rehabilitation (Relief Web 7 January 2024, Sio, 31 May 2024). The number of debris and rubble in the Gaza Strip stood at 37 million tonnes in mid-April, or 300 kilograms per square metre. On may Mungo Birch the head of the UNMAS program in the Palestinian territories said that:

> Gaza has more rubble than Ukraine and to put that in perspective, the Ukrainian front line is 600 miles [nearly 1,000 kilometres] long and Gaza is 25 miles [40 km] long. This rubble is likely to be heavily contaminated with UXO [unexploded ordnance], but its clearance will be further complicated by other hazards. There's estimated to be over 800,000 tonnes of asbestos, for instance, in the Gaza rubble. The sector as a whole will need hundreds of millions of US dollars over multiple years to make Gaza safe again for the population. (Middle East Monitor, 2nd May 2024)

This situation raised also questions about the scale, timeline and feasibility of temporary shelter erecting. Rafah governorate hosts over 1 million people in 65 km^2. Streets are filled with self-built structures and tents which are overcrowded and lack basic infrastructure. These temporary shelters constructed from salvaged materials were not resistant to cold, wet and windy winter weather. Insufficient shelter and inadequate nutrition increased risks of exposure and hypothermia and spread of infectious diseases. Moreover, evacuation orders, including those to collective centres, have caused fear and confusion as people are displaced to location with no real prospect of protection (Relief Web 7th January 2024).

It is worth noticing that the infrastructure consists of not only networks of material, technologically advanced structures and facilities, but also people—entities, groups and individuals who support functioning of these facilities and providing necessary services. This perspective bridging the gap between the business, societal and engineering views of critical infrastructures is applied by Choi et al. (2019) while describing the impact of Hurricane Michael and shelter management operations in Florida. Choi et al. (2019) identify primary infrastructure as "any entity that is committed to providing services essential for communities to sustain their economic and social activities in a normal situation". They are: municipal utility suppliers, hospitals and government agencies, county-owned resources and municipal services available to shelters for their operation and for the overall recovery of affected local communities. Whereas secondary infrastructure "consists of entities that are not expected to serve communities in a normal situation, but may emerge to support them when the primary infrastructure is not sufficient" They are: NGOs, local businesses and volunteer groups providing an external supply that are generally spontaneous and not committed to community recovery.

Due to the devastating impact of Hurricane Michael, 2018 which had destroyed buildings and infrastructure and generated an enormous volume of debris, the recovery for Bay County extended over time and many evacuees had to stay in hurricane shelters for longer as their houses were not safe enough. According to the US regulations the affected communities should be recovered to some extent before evacuees. The long-term operation of hurricane shelters is an element of the overall recovery process of affected communities. The last shelter closed on 30 of November, 2018, which was about 50 days after the first hurricane shelters opened (Choi et al., 2019).

Shelters during the all stages of disaster management were maintained by teams who service primary and secondary infrastructure in all main areas:

Information (on health, the use of backup generators, curfew policy, shelter operations and emergency reporting) was provided to the public by radio channels operated by public information officers (PIO). Whereas social media platforms (Facebook and Twitter) connected people and the PIOs. Traditional flyers, press releases and airplane message banners were also employed. Mobile companies (Verizon, AT&T and T-Mobile) provided ad hoc but reliable communication support to the shelters and the county even after the communication was restored. These companies used cells-on-wheels (COW) assets, whereas volunteers and disaster victims were provided with free mobile phones and mobile hotspots, supplementary systems such as lighting systems and gas generators charge the ad hoc communication equipment.

Failure in one of the pump stations and water treatment plants in Bay County caused health concerns and hindered the operation of the shelter facilities, as a result the shelter which did not have a backup means of wastewater disposal had to be closed and people relocated. The operators of public utility services restored power in collaboration with the secondary infrastructure utility companies from outside the county which deployed their repair crews. Till December 26, 2018, 99% of electricity service in Bay County had been restored.

Functioning of the shelter depends heavily on services provided to evacuees. Shelter is not merely a safe place for evacuees to stay, but it requires essential services such as the distribution of water, food and other necessities. Points of distribution (PODs) were opened to apportion emergency supplies. Social groups and organizations like the Salvation Army and the American Red Cross (ARC), World Central Kitchen (i.e. secondary infrastructure) also provided feeding services to affected communities: operated mobile feeding vehicles throughout Bay County and roved vehicles to reach those who could not access feeding sites, opened several feeding locations and kitchens (Choi et al., 2019).

There are also bunkers built for military purposes. The border guard tower is designed for security and monitoring, surveillance of border areas and used by military forces, border guards and other services responsible for border protection. WOT border mobile shelter is to protect the health and life of military personnel and civilians in the event of war or other threat. Whereas a concrete bunkers are for protecting military personnel and civilians during a war or non-military crisis in case of a shock wave resulting from an explosion of military artillery, improvised explosives or nuclear explosion. These kinds of bunkers have a base of a steel cube with double walls made of 6 mm thick steel and reinforcing ribs. After assembling at the site, the walls should be filled with C30/37 concrete (European Shelters).

For instance, in Bakhmut Ukrainian soldiers seek shelter in an underground bunker to avoid Russian drones constantly patrolling the skies (EuroNews, 2024, January 1).

Under the Steel Front military initiative, steel underground command posts are built to protect personnel from the effects of shelling and enable the Ukrainian Armed Forces to operate. Made of corrugated steel the headquarters consists of five or six separate mobile bunkers—interconnected underground shelters designed for various functions, from living quarters and sanitary areas to the command's working areas as they are equipped with communication systems, storage for weapons and personal belongings and electronic equipment (GMK Center; more in Proszowski et al., 2025).

2.6 Building Resilience

These days cities are challenged by variety of hazards and risks. Smart infrastructure designed to build urban resilience and reduce risks paradoxically expose urban systems to new forms of vulnerabilities. Infrastructure deterioration, overload or failure of critical physical and digital infrastructure or services lead to the breakdown of internet, cellular devices, public utilities or satellites. And all that may be triggered by cyberattacks, intentional or unintentional physical damage, or solar storms (WEF, 2023, Appendix).

Collapse or lack of public infrastructure and services as well as housing may be caused also by natural disasters, which emphasizes the need for greater care and investment in resilient buildings and critical infrastructure. The earthquakes in Türkiye 2023 caused more than 50,000 deaths, widespread demolition of buildings and over $34 billion worth total damage (4% of Türkiye's 2021 GDP). The majority of the damage was sustained by the housing sector. Türkiye's earthquake highlighted the impact of soil types on building integrity, therefore building resilient housing infrastructure requires knowledge on local geological conditions. Similarly, Japan, struck by earthquake on New Year's Day 2024, demonstrated the importance of planning and resilient building strategies. The 7.6 magnitude earthquakes caused the tragic loss of over 200 lives and more than 300 homes destroyed (World Bank, 2024).

Flooding, mostly a consequence of climatic change may be intensified by progressing urban expansion as drainage systems flow capacity is not sufficient and the infrastructure systems are maintained in a selective way.

> [...] cities were not designed for our changing climate. In addition, factors such as urbanisation, outdated infrastructure and geographical location significantly influence exposure to climate risks. [...] local governments predominantly prioritize technical infrastructure (e.g. transport and electricity systems) over social infrastructure like hospitals, nurseries and elderly care facilities. Surprisingly, critical elements like banks, data centres and crisis centres are often overlooked yet they are the backbone of our financial system. Infrastructure systems are interdependent, so failure of one system

> can cause other systems to fail as well. (Urban Insight by Sweco—Floods and critical infrastructure, p. 3)

Other consequences of the combination of climate change, urbanization and connectedness is a rise in infectious disease. The research confirms the relation between slum settlements and an increase in risk of insect-borne disease, malaria. Infectious diseases are more prevalent in urban areas and seasonal flooding. Mobility and interactions of residents of over-populated urban areas make local approaches to disease prevention and response ineffective (Kilcullen, 2015). Since these cascade effects may have unexpected and manifold societal consequences (disrupting vital societal functions, economic activities, public health and safety, or the environment), it is crucial to address also psychological impacts (stress and fear) the infrastructure failures may have on affected populations.

In response to these challenges, the European Union adopted the Critical Entities Resilience (CER) directive (2023, January 16). It identifies the critical entities for the sectors which provide essential services in upholding key societal functions and makes Member States responsible for risk assessments, identification of the critical entities and enhancement of their resilience.

List of essential services includes:

1. **Energy sector**, with services such as the electricity production and energy storage;
2. **Transport sector**, with services such as management and maintenance of airport or railways infrastructure;
3. **Banking sector**, with essential services such as taking deposits and lending;
4. **Financial market**infrastructure **sector,** with services such as the operation of trading venue and of clearing systems;
5. **Health sector**, with distribution, manufacturing, provision of healthcare and medical services;
6. **Drinkingwater sector**, with drinking water supply and drinking water distribution;
7. **Waste**water **sector**, with waste water collection, treatment and disposal services;

8. **Digitalinfrastructure sector**, with services such as the provision and operation of internet exchange point service, domain name system, top-level domain, cloud computing and data centre;
9. **Public administration sector** services;
10. **Space sector**, with the operation of ground-based infrastructure services;
11. **Production, processing and distribution of food sector**, with the large-scale industrial food production and processing, food supply chain services and food wholesale distribution services.

The extend the climate risk relates to critical infrastructure depends on three factors (Urban Insight by Sweco—Floods and critical infrastructure):

- The climate hazard and its chance of occurring, e.g. rainfall occurs and exposes a certain area to floods more frequently and heavily than in other locations.
- The vulnerability of the assets, e.g. whether a flood causes the failure of an asset to operate?
- The impact of any failure. Who or what will be impacted if the asset fails—society, the environment and/or the economy.

Thus, spatial planning and the planning of critical infrastructure needs to take into account the natural constraints of certain locations (e.g. flood-prone areas, weak soils, erosion) and technical solutions that could strengthen the systems resilience. There are variety of flood prevention methods applied. One of them is described in the literature as the "keeping water away from people". Dykes, floodwalls, embankments, dams, reservoirs and relief channels protect a specific area and hydrotechnical equipment from flooding. This approach is criticized as it gives a false sense of security created by protective structures and is also described as "the vicious circle of flood protection". If a given area is protected, for instance by creating dykes, it develops. When this area is flooded during the next flood, the losses exert more intense social pressure on modernization and new investments in resilient solutions (e.g. increasing the height of dykes). This cycle repeats. The "keeping the water where it falls" approach supports natural processes for retaining water in the landscape, for example, in swamps, peat bogs and forests, riverbeds and valleys.

The projects include: moving dykes away from riverbeds and reducing the high-water levels, or regulatory works like renaturization of rivers and constructing flow polders, dry reservoirs and lateral impounding reservoirs (Morawski, 2022).

Floods cause great damage to energy networks and breaks in energy supplies. Water and mud in the flooded equipment incurs costs of time-consuming reparations. Landslides and soil erosion are additional causes of damage to the power transmission grid. Moreover, flooding causes breaks in energy supplies not only in areas directly affected as electrical substations are located also in distant areas. Relocation of its vulnerable elements to areas above the expected flood level is a common practice. Discussing the resilience of the transport system to flood risk, one should consider not only all the possible damage to the road infrastructure and the cost of repairing it, but also the disruptions in routing and scheduling. Bridges are the weakest element as they and their access routes are easily destroyed. The transport system including roads and railway infrastructure are also vulnerable to damage. It may be concluded that urban areas can be mostly affected by floods. However, transport infrastructure will always run also through floodplains, which are parts of transport corridors. Thus, the practical approach will be to develop transport infrastructure and create its new elements, which may be more resilient to flooding (Morawski, 2022).

The infrastructure of water supply, both drinking and industrial water is also highly vulnerable to flooding:

> During every flood, microbiological, physical and chemical parameters of water are changed. The flood water might contain wastewater or agricultural fertilizers washed out. River ecosystems are imbalanced. Reservoirs are particularly at risk. The matter carried by a river is accumulated in the reservoir. Different types of substances, such as mineral deposits and organic sediments, or municipal waste are being introduced to it via flood water. It has particularly negative consequences for the water supply when a reservoir is the source of water supply for the population. (Morawski, 2022, p. 117)

The construction of flood protection infrastructure is of great economic importance. The constructed dry retention basin in Racibórz would reduce insured losses by 19% in 1997 flood and 11% in 2010 flood (Polish Chamber of Insurance, 2023).

Economic impact of floods increases material losses of households, companies, agriculture and the infrastructure in general. Consequently, the significant losses in insurance sector will lead to higher reinsurance rates. Costs of flood depend heavily on the wealth of the affected area. For instance, the flood similar in magnitude to the flood 11 years ago would cost 16. 2 billion PLN. It is by 3.4 billion PLN—20.9% more than in 2010. The increase is generated by the higher value of private and public assets on the threatened areas (Forsal.pl).

The list of challenges cities face is unquestionably lengthened by environmental risks like air pollution. As urbanization increases the air quality deteriorates. Cities worldwide are increasingly taking action to counter air pollution and address emissions to protect public health is critical. 50 cities participate in C40's Cities Clean Air Accelerator and 35 cities in C40's Green and Healthy Streets Accelerator and launch innovative strategies and initiatives to clean their air and protect the wellbeing of residents.

C40 Clean Air Accelerator signatories are taking an integrated approach to reduce air pollution with actions tackling emissions across a range of sectors (Clean Air Fund):

1. **Boosting air quality monitoring**

For instance, in September 2022, Bogotá launched their collaborative network of air quality microsensors, which will collect valuable information on top of their well-established air quality monitoring network.

2. **Implementing Low Emission Zones**

These zones give public space back to residents. Active mobility and sustainable transport options improve air quality and create more liveable communities.

Cities like London introduced the Ultra Low Emission Zone (ULEZ) which has contributed to an almost 50% reduction in toxic Nitrogen Dioxide pollution in central London. The ULEZ are to be expanded across all London boroughs on 29 August 2023, bringing clean air to 5 million more people.

3. **Enabling active mobility and greening spaces**

Many cities are expanding their cycle infrastructure: cycle lanes in Buenos Aires now total 267 km, while Paris is implementing a new bicycle plan to become a 100% bikeable city by 2026.

Cities are also expanding green cover in cities, for example Sydney's Greening Strategy and the city of Guadalajara that planted 22,000 endemic trees in 2022.

4. **Electrifying public transport**

Expansion and electrification of mass transit also reduces air pollution in cities. Bengaluru continues to extend its metro network and has increased the number of electric buses with 75 already operating and another 921 coming. Delhi is also currently working to electrify 80% of its bus fleet. The city already has 300 electric buses and is planning to introduce 8,000 electric buses by 2025.

5. **Phasing out dirty fuels for warming**

Warsaw has established a ban on burning coal to take effect in 2023, while Seoul and Washington D.C. implemented emissions standards for boilers. Cites are also phasing out old boilers:

Warsaw eliminated 1,225 classless and low-class stoves.

Seoul deployed 61,000 domestic and 487 commercially certified "eco-friendly" boilers.

Madrid, through the Change 360 Plan, is investing €14 million between 2021 and 2023 to renew old diesel and carbon boilers.

Air pollution is the number one environmental health problem in the European Union which is addressed by comprehensive clean air policy based on three pillars: ambient air quality standards, reducing air pollution emissions and emissions standards for key sources of pollution. Thus, way EU member-states should achieve the EU's zero pollution vision for 2050 (https://commission.europa.eu/eu-regional-and-urban-development/topics/cities-and-urban-development/priority-themes-eu-cities/air-quality-cities_en).

The crisis which is deepening in growing cities, is the result of little urban affordability. It is widespread and multifaceted as it reflects "non-existence, or widespread bankruptcy of social security systems and erosion of social security benefits, alongside inequitable or insufficient public infrastructure and services [...] affordable and adequate housing, public education, child and elder care, healthcare, transportation systems" (WEF, 2024).

This crisis deepens systemic injustice and inequities by pushing vulnerable populations, including low-income households, migrants, youth, elderly and minorities, into insecure living conditions. The number of people living in slums or other informal settlements has grown by 165 million in the past 20 years, bringing the total to nearly 1.1 billion, e.g. Freetown, Sierra Leone—one of the most crowded cities in the world. People construct their informal settlements on precarious land vulnerable to flooding, fires and landslides. Urbanization with unplanned housing development and deforestation were identified as the main causes of the tragedy in 2017, when more than 1,000 people lost their lives in a devastating landslide (https://www.aljazeera.com/news/2017/8/28/sierra-leone-mudslides-kill-more-than-1000).

In cities with open market economies, homes and small business premises are becoming increasingly unaffordable due to rapid population growth, profit-maximizing property investment, public authorities' incapacity to balance growth and build affordable housing or workspace. London may serve as the case. London's population has boomed over the past 40 years from 6.9 million in 1980 to 9.6 million today and London's small business base has doubled from around 500,000 to 1 million. London hosts 6 million jobs, £1.5 trillion housing assets and a £4 trillion property asset base. This trend leads to an enormous increase in housing prices and the costs of small business space.

Location is other factor creating the dilemma of living near job opportunities or at the periphery of cities especially for low-income households. Near areas with many jobs, there are the slopes on which favelas grew near Rio de Janeiro's centre. Whereas at the peripheries of cities, there is more affordable land and housing but more significant transportation costs, e.g.: the peri-urban fringe of Johannesburg and Mexico City. Such choices are incredibly challenging, especially for economic migrants (WEF, 2024).

Improving health and health equity for vulnerable populations requires addressing the whole set of health determinants. They include a wide range of factors in the political, economic, physical and social environment of health. Medical care accounts for 10–20% of health outcomes while social determinants like education and income account for the remaining 80–90%. Poverty is considered the most explicit determinant of health equity with cross-cutting impacts. New city- dwellers living within informal settlements are often those who lack decent work, access to safe and durable housing, clean energy sources, improved drinking water and sanitation (WHO).

Whereas affordability is a critical factor, accessibility within an urban context goes beyond this engaging many sectors to make city more inclusive. Accessible cities are designed and built in the way to offer intentionally inclusive infrastructure for their vulnerable people—the elderly who account for over 12% of the global population and the disabled, who account for one in six people globally. The UK port city of Bristol is characterized by incredible concentrations of both wealth and poverty. Marvil Rees the major of Bristol, UK explains how Bristol not always meets the expectations as accessible and affordable city:

> Some neighbourhoods offer the highest quality of life in the UK, but others are among the poorest in the country. In one area, Clifton, nearly every young person goes on to higher education; in Hartcliffe in the south, fewer than one in 12 do. People from around the country flock to Bristol to work in a thriving economy, but the city has been identified as one of the worst in the UK for racial inequality in education and employment. When local government seeks to make Bristol better, it is not working at the level of a single city, but a complex patchwork of poverty and prosperity. Just keeping track of this ever-changing landscape is a huge challenge – one that a current revolution in big data promises to address. (Ratti & Rees, 2023, October 19)

Rapid population growth in disaster-prone regions concentrates people where recovery after extreme events is costly. Disaster and climate risks in the built environment decrease property values, increase insurance premiums and compound recovery costs, impacting developers, financial institutions and local governments.

2.7 Conclusion

The technological, economic and social changes of the late twentieth century have made the city spread out, attract more people, which has increased its potential as well as vulnerabilities to the whole spectrum of threats. The growth of population means the growth of demand for resources. Infrastructure as a supply system has become a critical asset the dwellers are dependent on to the extent that any disruptions in the system make them completely powerless and lead to or intensify the existing crises or conflicts. This human vulnerability is exploited by variety of adversary groups (states as well as militias). The scarce and essential for life resources (water, energy) and their infrastructure may be tools or a target in conflicts applied for a political, tactical or psychological effect. Whereas natural disasters like floods destroying infrastructure cause health (infections spread), social and economic insecurities (unemployment, business collapse). Similarly, housing unaffordability and low or no sanitary systems resulting from natural disaster or weak governance may have multifold impact on population (infections, social inequalities, social unrests) which can easily be politically exploited by state and no-state actors (terrorists, gangs). Adaptive countermeasures high-tech solutions sensitive to geographical conditions, comprehensive policies supporting resource and infrastructure management and security projects and procedures for civil protection (building and maintaining shelters).

The discussed examples prove that all the threats are to some extend unpredictable, uncontrollable and our countermeasures, although continuously adapted to the city ecosystem conditions, leave the gap for unknown scenarios.

References

Agility Effect. (2019). *Communication Infrastructure for the Smart City*. https://www.theagilityeffect.com/en/article/communication-infrastructure-for-the-smart-city/ (3 June 2024).

Bilgiç, E., & Baba, A. (2023). Effect of Urbanization on Water Resources: Challenges and Prospects. In S. Ali & A. M. Armanuos (Eds.), *Groundwater in Arid and Semi-Arid Areas. Earth and Environmental Sciences Library*. Springer. https://doi.org/10.1007/978-3-031-43348-1_4.

Bowen, K. J. et al. (2022). *The Risk of Cascading Climate Change Shocks and Stressors* (Science for Adaptation Policy Brief 5). The World Adaptation Science Programme (WASP) Secretariat, UNEP, Nairobi.

Bogarov, H. (2024, April 26). *Kyiv1 Posted a Map of Shelters in the City of Kyiv.* https://kyiv1.com/en/kyiv1-rozmistiv-mapu-ukrittiv-mista-kiieva/

Choi, J., Sanusi, F., Ulak, M. B., Ozguven, E. E., & Abichou, T. (2019). *Significance of Secondary Infrastructure for Shelter Management in the Aftermath of Hurricane Michael* (Natural Hazards Center Quick Response Research Report Series, Report 295). Natural Hazards Center, University of Colorado Boulder. https://hazards.colorado.edu/quick-response-report/significance-of-secondary-infrastructure-for-shelter-management-in-the-aftermath-of-hurricane-michael

Cimino, S. (2017, December 7). *Defined by Our Infrastructure.* AIA Architect. https://www.architectmagazine.com/aia-architect/aiafeature/defined-by-our-infrastructure_o (2 June 2024).

Clean Air Fund. https://www.cleanairfund.org/news-item/5-ways-cities-are-cleaning-the-air-we-breathe/

CORDIS. *Security for Smart and Safe Cities, Including for Public Spaces.* https://cordis.europa.eu/programme/id/H2020_SU-INFRA02-2019/pl

Daoudy, M. (2020, September). Water Weaponization in the Syrian Conflict: Strategies of Domination and Cooperation. *International Affairs, 96*(5), 1347–1366. https://doi.org/10.1093/ia/iiaa131

EuroNews. (2024, January 1). https://www.euronews.com/2024/01/12/ukrainian-soldiers-face-fear-and-freezing-temperatures-in-bunkers

European Shelters. https://europeanshelters.pl/en/shelters/

Forsal.pl. (2021, September 11). *Katastrofy pogodowe to miliardowe koszty dla asekuratorów.* https://forsal.pl/finanse/artykuly/8239247,katastrofy-pogodowe-to-miliardowe-koszty-dla-asekuratorow.html

FORT WORTH, TEXAS CODE OF ORDINANCES; 2024 S-16 Supplement contains: Local legislation current through Ordinance 27032-06-2024, effective 6-25-2024, including Ordinance 27030-06-2024, effective 7-23-2024; American Legal Publishing 525 Vine Street, Suite 310 Cincinnati, Ohio 45202, Tel: (800) 445-5588, Internet: http://www.amlegal.com

Friedmann, J. (2005, October). Globalization and the Emerging Culture of Planning. *Progress in Planning*, 183–234. https://doi.org/10.1016/j.progress.2005.05.001

Gazeta Prawna. (2021, October 5). *Kryzys energetyczny w Europie obejmuje Północ.* Powód: niedobór wody. https://biznesiklimat.gazetaprawna.pl/energetyka-i-przemysl/8263427,Kryzys-energetyczny-Skandynawia-niedobor-wody.html

Gdańsk Tech. (2022, September 23). *New Groundbreaking System for Removal of Contaminants in Water.* https://pg.edu.pl/en/sustainable-development/2022-09/new-groundbreaking-system-removal-contaminants-water

GMK Center. (2024, January 20). https://gmk.center/en/news/metinvest-starts-production-of-steel-bunker-headquarters-for-the-armed-forces-of-ukraine/

Grelowska, M., Hibner, M., Deneka, W., & Jarynowski, A. (2024). *Information Flow During Culmination of Online Public Discourse Based on the Oder River Environmental Disaster.* https://www.researchgate.net/publication/382305905_Information_flow_during_culmination_of_online_public_discourse_based_on_the_Oder_river_environmental_disaster_accepted_in_E-methodology

Haidine, A., El Hassani, S., Aqqal, A., & El Hannani, A. (2016). The Role of Communication Technologies in Building Future Smart Cities. In I. N. Da Silva & A. R. Flauzino (Eds.). Smart Cities Technologies. InTech. https://doi.org/10.5772/64732

He, C., Liu, Z., Wu, J. et al. (2021). Future global urban water scarcity and potential solutions. *Nature Communications, 12*, 4667. https://doi.org/10.1038/s41467-021-25026-3

Inżynieria.com. (2023, April 4). *Mazowsze – w Ząbkach wybudują stację uzdatniania wody.* https://inzynieria.com/wodkan/wiadomosci/66349,mazowsze-w-zabkach-wybuduja-stacje-uzdatniania-wody © inzynieria.com

Jarynowski, A. (2022). *Oder River Disaster in Media, Infodemics.* Freie Universitat. https://blogs.fu-berlin.de/infodemic-corona/2022/08/19/oder-river-disaster-in-media/

KGPSP. https://www.gov.pl/web/kgpsp/konferencja-prasowa-dotyczaca-inwentaryzacji-budynkow-i-obiektow-budowlanych-przeprowadzonej-przez-funkcjonariuszy-panstwowej-strazy-pozarnej

Kilcullen, D. (2025). *Out of the mountains: The coming age of the Urban Guerrilla.* C Hurst & Co Publishers Ltd.

Kowalkowski, S., Kaźmierczak, D., & Paul, S. (2024). *Civil protection systems and disaster governance: A cross-regional approach.*

Liotta, P. H., & Bilgin, A. (2014). Why Human Security? The Case for Resilience in the Urban Century. In M. Martin & T. Owen (Eds.), *Routledge handbook of human security*. Routledge.

Maintained HCR/MWI/2021/003.

Marchal, H., & Stebé, J.-M. (2014). Communication Is Inherent to the New Urban Condition. *Questions de communication* [En ligne], 25 |, mis en ligne le 01 février 2014, consulté le 03 juin 2024. http://journals.openedition.org/questionsdecommunication/9494, https://doi.org/10.4000/questionsdecommunication.9494.

Middle East Monitor. (2024, May 2). *UN Says Scale of Destruction in Gaza Much Greater Than in Ukraine.* https://www.middleeastmonitor.com/20240502-un-says-scale-of-destruction-in-gaza-much-greater-than-in-ukraine/

Moody, J., & White, D. R. (2003). Structural Cohesion and Embeddedness: A Hierarchical Concept of Social Groups. *American Sociological Review, 68*(1), 103–127. 22 April 2023. https://www.researchgate.net/publication/233729906_White_D_Structural_Cohesion_and_Embeddedness_A_Hierarchical_Concept_of_Social_Groups_American_Sociological_Review_68_103-127

Morawski, M. (2022). Flood Vulnerability of the Critical Infrastructure in Poland. *Security and Defence Quarterly, 39*(3), 108–122. https://doi.org/10.35467/sdq/150456.

NIK. (2018). *Marna woda rury rwie*. https://www.nik.gov.pl/najnowsze-informacje-o-wynikach-kontroli/nik-o-utrzymaniu-i-eksploatacji-sieci-wodociagowych-w-miastach.html. 29 November 2025.

Notes from Poland. (2024, March 13). *Warsaw to Spend €27 Million on Emergency Preparedness, Including Shelters*. https://notesfrompoland.com/2024/03/13/warsaw-to-spend-e27-million-on-emergency-readiness-including-shelters/

Polish Chamber of Insurance. (2023). A Climate of Mounting Losses.

Proszowski, A., Wenski, J., & Niwa, P. (2025). BIULETYN. Inżynieryjne przygotowanie pozycji obronnych w świetle doświadczeń z konfliktu zbrojnego na Ukrainie Centrum Doktryn i Szkolenia Sił Zbrojnych, Bydgoszcz.

Ratti, C., & Rees, M. (2023, October 19). *How Big Data and Local Politics Can Make Our Cities More Inclusive*. https://www.weforum.org/agenda/2023/10/how-big-data-local-politics-cities-more-inclusive/ (30 June 2024).

Relief Web. (2024, January 7). *Gaza Crisis 2023—Shelter Situation Update*. https://reliefweb.int/report/occupied-palestinian-territory/gaza-crisis-2023-shelter-situation-update-7th-january-2024.

Schillinger, J, Özerol, G., Güven-Griemert, S., & Heldeweg, M. (2020). Water in War: Understanding.

Sio, M. (2024, May 31). Overcrowded Shelters in Gaza Fuel Spread of Diseases, Warns UN Agency.

State Emergency Service of Ukraine No 03-1870/ 162- 2, 14th June 2022.

Statista. https://www.statista.com/statistics/438840/investment-in-inland-waterway-transport-infrastructure-in-poland/

Szlak Podziemi. http://szlakpodziemi.pl/

Szpunar, M. (2008). Przestrzeń Internetu - nowy wymiar przestrzeni społecznej. In A. Siwik & L. H. Haber (Eds.), *Od robotnika do internauty. W kierunku społeczeństwa informacyjnego*. Akademia Górniczo Hutnicza, 9. 22 April 2023. http://www.magdalenaszpunar.com/_publikacje/2008/przestrzen_internetu.pdf

Tachi, T. et al. (2019). Overview and Examples of Water Infrastructure Solutions. *Hitachi Review, 68*(5).

Temet. (2020). *Critical Infrastructure*. https://www.temet.com/solutions-and-services/critical-infrastructure the impacts of armed conflict on water resources and their management. WIREs Water 7: e1480. https://doi.org/10.1002/wat2.1480

UN Partners Portal. (2024). Call for Expression of Interest Shelter and infrastructure established, improved and maintained. HCR/MWI/2021/003.

UN Water. Water and Urbanization. https://www.unwater.org/water-facts/water-and-urbanization
UNHCR Global Focus. https://reporting.unhcr.org/shelter-and-infrastructure-49
WEF. (2023). *The Global Risk Report 2023* (18th ed.).
WEF. (2024). *These Are Some of the World's Top Smart and Sustainable Cities*. https://www.weforum.org/agenda/2024/03/top-cities-smart-technology-sustainability/
Westman, L., Patterson, J., Macrorie, R., Orr, C. J., Ashcraft, C. M., Castán Broto, V., Dolan, D., Gupta, M., van der Heijden, J., Hickmann, T., Hobbins, R., Papin, M., Robin, E., Rosan, C., Torrens, J., & Webb, R. (2022, June). Compound urban crises. *Ambio, 51*(6), 1402–1415. https://doi.org/10.1007/s13280-021-01697-6. Epub 2022 Feb 14. PMID: 35157255; PMCID: PMC8853022.
WHO. *Environment, Climate Change and Health*. https://www.who.int/teams/environment-climate-change-and-health/healthy-urban-environments/urban-planning/health-equity
World Bank. (2024). *Global Earthquakes Underscore the Importance of Resilient Infrastructure*. https://www.weforum.org/agenda/2024/03/global-earthquakes-resilient-infrastructure/
World Bank Group. (2014, January 20). *Water Shortages Slow Energy Production Worldwide*. https://www.worldbank.org/en/news/press-release/2014/01/20/water-shortages-energy-production-worldwide
World Economic Forum. (2024a, March 18). *Air Pollution Is a Silent Killer. Here's How Cities Are Tackling It*. https://www.weforum.org/agenda/2024/03/air-pollution-is-the-silent-killer-heres-how-cities-are-tackling-it/
World Economic Forum. (2024b). *Reshaping Affordability: Interventions for Inclusive and Livable Cities*. https://www3.weforum.org/docs/WEF_Reshaping_affordability_2024.pdf
World Resources Institute. (2015, August 26). *Ranking the World's Most Water-Stressed Countries in 2040*. https://www.wri.org/insights/ranking-worlds-most-water-stressed-countries-2040
Wysocki, K. Shelters in Ukraine; Presentation. Zakład Inżynierii Wojskowej i Maskowania, Instytut Wsparcia i Zabezpieczenia Działań Wydział Wojskowy, Akademia Sztuki Wojennej, k.wysocki@akademia.mil.pl.
Wysota. (2023, September 11). Polsce grozi potężny kryzys. Woda może pogrążyć gospodarkę. Money.pl. https://www.money.pl/gospodarka/eksperci-bija-na-alarm-woda-moze-pograzyc-nasza-gospodarke-6938745722481280a.html
Xie, B., He, D., Mercer, T., Wang, Y., Wu, D., Fleischmann, K .R., Zhang, Y., Yoder, L. H., Stephens, K. K., Mackert, M., & Lee, M. K. (2020, December). Global health crises are also information crises: A call to action. *Journal of*

the Association for Information Science and Technology, 71(12), 1419–1423. https://doi.org/10.1002/asi.24357. Epub 2020 Mar 13. PMID: 32427189; PMCID: PMC7228248.

Yang, S., Fichman, P., Zhu, X., Sanfilippo, M., Li, S., & Fleischmann, K. R. (2020). The Use of ICT During COVID-19. *Proceedings of the Association for Information Science and Technology. Association for Information Science and Technology, 57*(1), e297. https://doi.org/10.1002/pra2.297.

Zawahri, N.A., & Weinthal, E. (2019). The intersecting dimensions of water security. In M. Gueldry, G. Gokcek, & L. Hebron (Eds.), *Understanding new security threats*. Routledge.

ZBiAM. (2022, June 2). *Jaśmin Also for Crisis Management*. https://zbiam.pl/jasmin-tez-dla-zarzadzania-kryzysowego/

Zeihan, P. (2022). *The end of the world is just the beginning: Mapping the collapse of globalization—The Collapse of globalization and its aftermath*. Harper Business.

CHAPTER 3

Urban Warfare Dimensions

Abstract People are rather well educated and organized to face simple disasters, yet, they are much less competent when facing complex risks in the interconnected systems that build the structures of our world, such as organizations, economies, societies and the environment. Thus, effective preparedness requires building social capacity through education and training. The chapter present the design thinking as the teaching approached tailored for the security needs. The theoretical theories supporting this choice as well as methodologies and practical implementation in short/agile courses are presented and skills frameworks analysed as the guidelines for security education and training programs.

Keywords Design thinking · Agile project management · Capacity building · Teaching/learning theories

3.1 Cities as a Battlefield

Urban warfare especially city defence is one of the oldest, best known and complex concepts of the art of war and military operations in urbanized areas have always been a challenge for the armies. The defensive walls built around cities and castles have always been fortifications that the advancing army have had to made a lot of efforts to destroy them.

S. Kowalkowski et al., *Cities in Crisis*,
https://doi.org/10.1007/978-3-032-08843-7_3

It is worth starting the historical review of the urban combat with the conquest of the future Promised Land by the Israeli army commanded by Joshua. One of the targets was Galilee. Kings of the Northern Palestine led by Jabin, king of Hazor stood against the Joshua's troops. The battle was waged at Merom river. After winning a victory, Joshua continued conquering Hazor and other cities of the northern Palestine.[1]

Cities were also a challenge for the Alexader of Macedon who in 334 B.C. commence the great campaign against Persia Empire of Darius III. Alexanders' army defeated the adversary at Issos (333 B.C.), Tyre[2] and eventually Gaza and Egypt. In 331 he decisively defeated Darius III at Gaugamela and incorporated Mesopotamia and the area of today Iran and Afghanistan into his growing empire.[3] Sine qua non of the victory was the conquest of the cities in the way of the advancing army. It is worth mentioning also that before the expeditions against the Persians, Alexander organized the campaign at the Balkans and defeated rebellious Thebes and other Greek city states.

Another great ancient leader, Julius Ceasar sieged Gauls in Alesia. Gauls's king, Vercingetorix after sending back the cavalry, locked himself

[1] According to the biblical chronology the Battle of Jericho as the part of the Joshua's conquest of the Promised Land took place in 1220–1200 B.C. Israeli army counted about 40,000 warriors. The existence of the ancient ruins of Neolith and Bronze is confirmed by the archaeological research, yet the destruction of the city is dated back to the times of the Egyptian conquests, a hundred years earlier than the Bible reads. Archaeologists have tried to find the traces of the battle for years, however, in spite of many attempts the ruins of the city from that period have not been found (Krawczyk, 2012).

[2] People of Tyre tried to defend the city relying on strong fortifications and advantageous geographical location (on the isle 500-meter far from land). Moreover, Tyre has a strong fleet which Alexander lacked. The Macedonians stated over half-of-the year long siege. During that time, they built a causeway joining Tyre with the land and prepared fleet of 200 warships which blocked the city from the sea. After gaining the advantage at sea the Macedonians started building causeways along city walls where they positioned siege engines and systematically damaged fortifications. After destroying the part of the defensive wall, they attacked first from the causeway and then from the warships. The city was defeated after the concentrated attacks from various directions (World History Encyclopedia. (2011, Aug.8). Alexander's Siege of Tyre, 332 BCE).

[3] The localisation of the Gaugamela Battle that took place before 2,300,000 years was identified on the territory of today Iraqi Kurdistan near Tell Gomel by Polish researchers. Being uncertain about the place of the battle, the researchers point to the area of Tell Gomel and Mosul. Michał Marciak's research team at Historical Department of Jagiellonian University finished the first season of explorations on the probable place of the Gaugamela Battle in 2018 (Jagiellonian Univeristy in Krakow, 2023).

in the city waiting for the support to arrive. Aware of the danger he ordered to build two line of fortifications—inner one to protect them against the attacks from the settlement and outer one, to protect them against the succour. Gauls' successive attacks at the Roman fortifications did not bring an expected result. As the Roman infrastructure was well-developed the Gauls did not even approach walls and had to focus on backfilling fosses. Eventually warriors from Alesia had no hope for success and capitulated (Krawczuk, 1990).

Likewise in the Middle Ages and in the later centuries, for instance in Poland of the Piast's and the Jagiellonians' the defensive capacity of the army depended on how well cities and castles were prepared. The defensive system built by Casimir III the Great guaranteed the enduring defensive capacities of Poland till seventeenth century. Due to centuries-old negligence of the national defence strategy cities declined. However, still in the seventeenth century during the Swedish Deluge the cities: Częstochowa, Zamość, Przemyśl and Lwów withstood and contributed to the overall victory. In 1974 also The Kościuszko Uprising serves as the example of the excellent defence of Warsaw against the Prussian and Russian troops. The war of 1920 proved the cities to be a key element in the defence system as well.

In seventeenth and eighteenth centuries although cities were not treated as fortresses any more, they were still points at which the advancing adversary could have been stopped. This happed in the Thirty Year's War (1618–1648), Napolean's wars, Crimean war or Civil war (1861–1865).

In the twentieth century the successful defence of Lwów stopped the Northern-Western Bolshevik Front and the heroic defence of Radzymin near Warsaw created the advantageous conditions for manoeuvre of Polish forces (they held back and engaged the enemy) and for counterattack on the Wieprz River where they pushed back the enemy forces. After WW I most states prepared the defence of cities as the element of the strategic border protection. Siegfryd Line[4] in Germany, Maginot Line in France and Molotov Line in USSR were built.[5]

[4] Siegfried Line—system of 15 560 pillboxes and strongpoints built along the German western frontier with France and Luxemburgh about 600 km long.

[5] The system consisted of thirteen Fortified Regions along the front from 80–100 km long. The first defuse line were the field fortifications called an obscuration area. In the distance of several or dozen kilometres inland there were so called defence nodes which

The partially successful defence of the Polish capital organized ad hoc by over 100 000 people (soldiers of regular Polish Army, voluntary formations and considerable number of civilians) forced Germans to the long-lasting efforts (1–27 September 1939). The enemy used 11 divisions of land forces supported by 2 air fleets (Odziemkowski, 1989).

The defence of Polish military depot at Westerplatte exemplifies how important well-developed fortifications for city defence were. The troops using the well-fortified and prepared depot area could defend themselves successfully for a long time with limited means and forces.

Ignoring the role of cities in the state defence strategy was the reason why the USSR lost the strategic initiative and war in summer 1941 (https://histmag.org/Stalingrad-wojna-szczurow-7563).

The defence of Stalingrad, the city on the Volga downstream made a history of the art of war. In this city every house was a fortress to fight for. The decimated defenders focused on the considerable forces of German Army Group South whereas at the back marshal Georgy Zhukov prepared counterattack at the weaker wing of the enemy. As the result of the Russian offensive actions The Red Army took over the initiative and Wehrmacht had to retreat. The Russian army regained almost all territories won by Germans during "Fall Blau" operation.

After WW II the role of cities and urbanized areas has not changed.

Urban battles are usually fought when (FM 90-10-1, 1982, p. 1–5):

- A city is between two natural obstacles and there is no bypass
- The seizure of a city contributes to achieving an overall objective
- The city is in the path of a general advance and cannot be surrounded
- Political or humanitarian concerns require the seizure or retention of a city.

The operations in the urbanized areas from Mogadishu, Kabul to Baghdad have become the new research area for the military research centres and institutions all over the world (The NATO Urban Warfare Project). The newly created laws and operational requirements for armed forces are to increase the effectiveness of urban combat (Makowiec &

consisted of the network of shelters for artillery and machine guns. The Fortified Regions included 8–9 defence nodes and 3–5 points of resistance. The system was completed by closely located and mutually protective shelters with obstacles for the infantry and anti-tank barriers. Molotov Line was to stop the enemy for the time necessary for mobilization and then to support offensive actions of the own troops.

Mroszczyk, 2012) by implementation of the innovative approaches and experiments with real-time modelling and simulation methods conducted in NATO (Urban Combat Advanced..., 2010). These innovations enhance preparation of the armed forces to urban conflicts as well.

Yet, the literature proposes two different stands on the nature of modern urban war. One is that the urban battles of today are not new and they have been seen before (Betz & Stanford-Tuck, 2019; Hills, 2004) whereas another view emphasizes that they are much different and this difference is not merely about weaponry: "Although certain features of urban combat endure, the physiology of urban warfare has changed" (King, 2021). Warfare is a social organization as it is influenced by social, economic, political and communication processes. It is about the interplay of social and physical topography, weaponry and force which creates highly complex urban environment and makes today urban war different or new (Danielsson, 2025, p. 405; King, 2021; Kilcullen, 2015; Weismann, 2024).

The factors accounting for complexity of the urban warfare environment are numerous. Although they are applicable for defining its main parameters, they will not enable to draw any scenarios of the future battles. The factors include: size of the armies, urbanization, globalization, technological advances, communication, climate change, population growth, littoralization, insurgency, (Weismann, 2024; King, 2021; Kilcullen, 2015). McCarthy (2018) identifies five metatrends as 'C5L' (crowded, connected, collective, constrained, constant and lethal). While analysing peculiarities of urban warfare we refer to all these processes (variables) as they are intertwined and produce synergic/compound impact.

However, the complex urban environment itself is neutral: "It affects all protagonists equally, even if it does not always appear to do so" (Betz and Stanford-Tuck, 2019, p. 61). Its complexity can be overcome by a combination of realistic hard training, changes in command mindset at the strategic, political and tactical levels, high tech and knowledge (Betz and Stanford-Tuck, 2019; Danielsson, 2025, p. 406). Yet, these levels should not be seen as hierarchically structured but rather highly interwoven (Betz and Stanford-Tuck, 2019).

3.1.1 *Method*

Considering the urban ecosystem (nature, infrastructure and people) as the warfare environment and above—shortly discussed approach to contemporary urban warfare of interrelated processes (variables) and command and control levels we organized the analysis of urban warfare in three mingling dimensions:

- physical one as the characteristics of urban landscape determine the nature of urban combat, tactics, the use of weapons and their results to some degree—we call it within the city;
- operational referring to the scale and type of operations designed with the specific processes, assets (high tech, knowledge, competencies), methods and procedures—we call it between the peace and war;
- social one in which society becomes a deliberate target or means of fight, which reflects ideas, strategies and policies the society believes in and materializes—we call it within the society;

The aim is to describe how the urban warfare is commanded and controlled with all the consequences incurred by civilians.

3.2 Within the City

The characteristics of urban landscape determine to some degree the nature of urban combat, the use of weapons and their results (ATTP 3-06.11). Moreover, the reduced size of today armies results in particular tactics. Inner-urban siege of the twenty-first century cities take the form of partial or temporary blockade and involves also encirclement and contravallation of enemy positions within the city:

> As Mosul has shown [inner-city sieges] involve bitter fighting at close quarters between troops, tanks, armoured vehicles and bulldozers. Yet, they are highly complex, involving conventional, hybrid, irregular forces on the ground and the aerial platforms flying high above the city. Precision artillery deployed miles outside the city, information and cyber operations, collaboration with local militias and civil agencies have all become critical to the outcome. Consequently, the urban battle has localized onto specific sites within the city, but it has simultaneously extended out across

the global urban archipelago by means of social media and information networks (King, 2021, p. 14).

The nearby walls enclose urban engagements and soldiers need to consider their weapon's effects: arming range, muzzle blast and back-blast, the round's effect on the target and the target's effect on the round as well as enemy weapons effects on friendly positions. The engagement time is also shortened. Enemy personnel and vehicles can rapidly hide behind the nearby wall and then fire from the same or different positions. The engagement is predominantly characterized by close combat: riflemen is expected to hit targets through bunker apertures, windows and loopholes.[6] This can be done with accurate semiautomatic weapons. About 90% of all targets are engaged within a range of 50 meters or less.

Building construction types are also detrimental. Over 60% of buildings in the world are built of 2- to 24-inch brick or concrete block. Most large buildings constructed after World War II are resilient to the blast effects of bomb and artillery attack. Generally, they burn easily but retain structural integrity, remain standing and can be damaged further only with substantial explosive force. Thus, they can still be exploited for individual firing or as unit positions. City structures provide ready-made strong defensive positions therefore soldiers should attack this structure before attacking enemy personnel inside. They need also select and use appropriate weapons and demolitions considering their effects against masonry and concrete. Urban structures have typically hard, flat, smooth surfaces thus, rounds usually impact at some angle of obliquity, which reduces the effects of a round and increases the threat of ricochets. The depression and elevation limits for some weapons create dead space. Tall buildings form deep canyons that inhibit the use of indirect fires. Small-arms weapons, for example, can be fired with the deliberate intent of the rounds ricocheting behind cover to inflict casualties. Moreover, smoke, dust, shadows and the lack of light penetrating inner rooms reduce visibility

[6] Man-size holes include (ATTP 3-06.11):

Loopholes—firing apertures (a minimum of 8 inches in diameter) made in a structure.

Mouseholes—openings made to the interior or exterior of a structure (walls, floors, ceilings, and roofs) to aid inter- and intra-building communications and movement. A mousehole is usually a minimum of 24 by 30 inches in size.

Breach Holes—openings made in a structure using mechanical, ballistic, explosive, or thermal means to aid the entry of assault elements. A breach hole is a minimum of 50 by 30 inches in size.

and increase a sense of isolation. Thus, targets (even those at close range) tend to be unclear. Urban fighting often becomes confused melees with several small units attacking on converging axes or being attacked from multiple directions—the risks from friendly fires and ricochets. The peculiarities of the urban operational environment draw armies into fighting, as the irregular forces/ non-state armed groups/ militias usually avoid facing their enemy in the open by intermingling with the civilians.

Technology as one of driving forces of urban combat development offers a wide range of solutions, equipment and weapons: Artificial Intelligence, natural language processing, computer vision, nano- and biotechnologies, advanced big data analytics speed the revolution in sensor technologies, long-range, hypersonic "intelligent" weaponry, "swarms of killing machines" and directed-energy weapons (Allen, 2024).

For more accuracy, the list of the latest weapons for urban warfare includes (European Security and Défense, 2022):

- precision-guided weapons with Non-Line-of-Sight (NLOS) guidance capability, e.g.: mortars that use a high-trajectory to hit the enemy in narrow streets and behind buildings;
- Long-range precision-guided weapons that deliver support fire from a distance;
- Vertical Take Off and Landing (VTOL) drones, for intelligence, surveillance and attack whereas for early warning and electronic combat GNSS jammers are used;
- Thermobaric weapons that create a lethal overpressure, programmable explosive grenades, delay-fused aerial bombs or projectiles are all used to penetrate walls, roofs or multiple storeys;
- Ultra-Wide-Band (UWB) Radar, flying drones, "legged, wheeled and tracked robots" are used for tracking objects (people) in buildings. Augmented reality can support early warning, mapping and clearance of building interiors or subterranean routes as well as overwatching.

Psychological warfare exploits the internet, harnessing mass communications and social media, live broadcast by reporters and ordinary smart citizens.

Social media is powerful enough to communicate by the witnessing citizens a real time (e.g. Russian-Ukrainian) war coverage and this way make mere mortals feel they are in the middle of the conflict, crossing geographical borders:

> Through the power of social media, we've witnessed civilians confronting soldiers, Ukrainian soldiers holding the line and apartment dwellers recording incoming mortar shells in the distance from a birds-eye perch. Notable social media posts include the official twitter account of Ukraine asking for cryptocurrency and Mykhailo Fedorov, the vice prime minister of Ukraine and minister of digital transformation, asking Elon Musk for more Starlink stations to keep the internet running. A world away, a student at the University of Central Florida is tracking the yacht locations of oligarchs in real time on his Twitter account. Once posted, these bits of information fly to all parts of the world. (Halloran, 2022)

ICT tools are not reserved for the regular forces as it is relatively easily accessible also for "the small size of dangerous, insurgent groupings" able to "spread mass disruption and cause mass destruction" (Allen et al., 2021, p. 33).

The defenders fight for support for their cause using these tools while the attackers' use them to deceive, demoralize the enemy and win without fighting. The digital domain creates not only an opportunity to use the records to target the enemy but also its infrastructure (Emerging military technologies applied to urban warfare, 2017) such as the electricity grid or transportation networks, banking institutions and health facilities.

Technologies are also used for humanitarian aid delivery with unmanned evacuation vehicles whereas thanks to limited troop contact and precision targeting they contribute to a reduction in both combatant and civilian casualties.

High-tech solutions that provide precision targeting may reduce incidental loss to non-military personnel and objects. However, several concerns have been raised. The physical, psychological and mechanical distance from a victim facilitates using lethal force and consequently increases the number of military and civilian casualties. Another concern is whether; by reducing the human cost, the use of such weapons could reduce incentives for resolution and result in a protracted conflict. There is also an open debate on the possibility of hacked robots performing indiscriminate attacks (Emerging military technologies applied to urban warfare, 2017).

Moreover, weapon systems originally designed for use in open battlefields are used against military objectives located in urban populated areas where their effects are indiscriminate and devastating for civilians and critical infrastructure (e.g. hospital facilities). The examples of the highly populated areas where civilians were affected by using explosives[7] include:

> **Iraq**, 2016—42 incidents involving the use of heavy explosive weapons in Fallujah, 37 of which were described as indirect fire of mortars and rockets. 115 civilians dead and 150 injured in all 42 incidents were reported.
> **Syria**—15.5 million people needed water, sanitation and hygiene services in 2019, partly because of heavy infrastructural damage from the use of heavy explosive weapons.
> **Yemen**, 2015—the destruction of Hayden Hospital in Sa'ada by air strikes left 200,000 people with no access to lifesaving medical aid.
> **Libya** (April and July 2019)—over 120,000 civilians were displaced mainly as a result of the use of heavy explosive weapons in residential areas of Tripoli.
> **Gaza**, 2009 children accounted for a third of all civilian casualties; of the 353 children who died, 82% were killed by heavy explosive weapons.
> As of 14 July 2014, a total of 1,539 Palestinian casualties including 178 deaths as the result of missile strikes were reported. UN estimates, 77% of total deaths were civilians.
> **Afghanistan**, 2020—over 2,000 civilian casualties were reported from indirect fire from mortars, artillery and rockets.

Donatella Rovera, Senior Crisis Response Adviser at Amnesty International states firmly:

> It is high time for states to commit to stop using explosive weapons designed decades ago for open battlefields – such as artillery, rockets and heavy, air-delivered bombs – in populated civilian areas. In the 21st

[7] Explosive weapons in populated areas (EWIPA) are a subset of what is commonly referred to as conventional weapons. Technologically speaking there are many types of explosive weapons, including bombs, artillery shells, grenades, missiles, landmines and improvised explosive devices, which are activated by detonation of a high explosive substance; UN, Explosive weapons in populated areas, https://disarmament.unoda.org/convarms/explosive-weapons/ (July 31, 2024).

century, there is no excuse to use explosive weapons haphazardly near civilians. Militaries often have much more accurate and precise munitions at their disposal but use cheaper options instead. This may cut costs for the fighting forces but fundamentally it's civilians who pay the ultimate price, with their lives". —The Amnesty International (2021) webpage reads

Urban combat may result in large-scale destruction to buildings and consequently bring many casualties. Survivors, both military and civilian who are seriously trapped in the rubble can be extracted only with heavy construction equipment and rescuing can cause even more casualties as the rubble shifts and collapses on victims. Evacuation may also cause additional injury as the rescuers have to face numerous obstacles. Fires which are usually uncontrollable and explosions generate significant amounts of dust to the atmosphere; even the rescue activities themselves create dust, which limits visibility considerably. Moreover, the transportation of people, supplies and goods is disrupted: interrupted flow of traffic, destroyed roads, ports, bridges and rail lines. Although most city-dwellers escape from the city some remain behind close the area of the fighting and are, thus, in danger (ATTP 3-06.11, Appendix B).

On 18th November 2022 *the Political Declaration on Strengthening the Protection of Civilians from the Humanitarian Consequences arising from the use of Explosive Weapons in Populated Areas* was formally adopted by 83 states. It is the first, landmark document committing states, to "implement and, where necessary, review, develop or improve national policy and practice [...] to help avoid civilian harm, including, by restricting or refraining as appropriate from the use of explosive weapons in populated areas, when their use may be expected to cause harm to civilians or civilian objects. The declaration may also contribute significantly to strengthening respect for International Humanitarian Law (IHL) (Bagsaw, 2023) by among others, ensuring "comprehensive training of our armed forces on the application of International Humanitarian Law and on the policies and good practices to be applied during the conduct of hostilities in populated areas to protect civilians and civilian objects" (The Political, Section 3, 3.2.).

The implementation of the Political Declaration..." is strongly promoted by the UN Secretary-General in his *New Agenda for Peace*, where he recommends that states:

> Strengthen protection of civilians in populated areas in conflict zones, take combat out of urban areas altogether, including through the implementation of the Political Declaration on Strengthening the Protection of Civilians from the Humanitarian Consequences Arising from the Use of Explosive Weapons in Populated Areas, adopted on 18 November 2022 and establish mechanisms to mitigate and investigate harm to civilians and ensure accountability of perpetrators. (Our Common Agenda Policy Brief 9: A New Agenda For Peace, p. 21)

These international laws and declarations are the perfect exemplifications of evolving global governance as the mechanism for addressing the shocks, crises and conflicts single states are not able to deal with: "The rules and laws that apply to individuals as well as states are being negotiated among the family of individuals, groups and institutions [...] This is why it is possible to talk about blurring the distinction between inside and outside, domestic and external." (Kaldor, 2014, p. 68).

3.3 Between War and Peace

Technology and globalization have increased interconnectivity and interactions between states and non-state actors who "are also equally capable of seeping through borders, both physically and virtually" (Maskan, 2021) and turning fragile international systems into which Maskan calls globalized "interdependent anarchy" (p. 45). Al-Qaeda incited the Global war on Terror attacking Washington in 200 whereas in 2008 small group of bankers caused the worst financial economic and financial crash since 1929. Global interdependency enables all parties to share military technology, influence decisions and feed conflicts in technologically advanced and networked (smart) cities. Russia supporting the Syrian regime, bought 'Forpost' drones[8] from Israel, enabled Assad to stay in power and even decided which places to bomb and when. The US, Turkey and Iran also responded and interfered in the internal affairs of the country, giving support allied with their national interests and as a consequence prolonged the civil war (Laub, 2023).

[8] Forpost-R, the Russian version of the Israeli Searcher Mk 2 reconnaissance drone, produced under license in Russia with the addition of air-dropped bombs, used also as an Unmanned Combat Arial Vehicle (UCAV). The Forpost drone were used during Russian military actions in Syria and the invasion to Ukraine 2022; https://cat-uxo.com/explosive-hazards/ucav/forpost-ucav (August 1, 2024).

Technology increases the interconnectivity also directly at the urban battlefield which is no longer geographically limited.[9]

> There is simply no such thing as 'outside the battle space'. You need to be prepared to defend against and counter a wide range of hybrid attacks, kinetic as well as non-kinetic, hybrid threats and influence operations everywhere, including in yours and your partners' home country. (Weismann, 2023, p. 150)

These types of conflicts evolve over time and move from one city district to another, change from peaceful coexistence, through irregular warfare, to traditional wars between states. They are difficult to predict and vary in intensity, quality, capacities exploited and are played out in gray zones. The gray zone does not refer to geographical location but conceptual and operational space between peace and conventional war. Notions of risk and response are crucial as operations are conducted below the level that would mandate a conventional military response. The examples of low-profile activities which are calibrated or sliced (salami-slicing tactic) to avoid a substantial, kinetic response from the affected party on one hand and on another to gain expected benefits are: a private Chinese oil rig anchoring inside Vietnam's exclusive economic zone, a Chinese frigate chasing off a Philippines survey ship over Reed Bank, or a Chinese infantry platoon on a pile of rocks near the Spratly Islands (Jordan, 2020). The example of the failure of salami tactics is the incident of the Turkish Air Force shotting down a Russian fighter jet in 2015 for violating Turkish airspace after ten warnings to change course over the course of five minutes. The Russians considered it below the threshold

[9] "Multi-Domain Operations refers to the NATO concept to orchestrate military activities across all operating domains and environments. These actions are synchronized with non-military activities and enable the Alliance to create desired outcomes at the right time and place—to deliver a streamlined defence and deterrence system to its one billion citizens. This will enable NATO's Military Instrument of Power to prepare, plan, orchestrate and execute coordinated activities in collaboration with other stakeholders and actors associated with the Alliance. In doing so, this concept will deliver tailored options that build an advantage in shaping, contesting and fighting and presents dilemmas that decisively influence the attitudes and behaviours of adversaries. The key differentiating factor between "Joint" and "Multi-Domain" Operations is that "Joint" command structures promote coordination with other armed services, while a "Multi-Domain" mindset must go beyond that to include military and non-military assets, which is. Multi-Domain Operations in NATO—Explained October 5, 2023; https://www.act.nato.int/article/mdo-in-nato-explained/

while Turkey repeated warnings and finally acted upon the situation to preserve the status quo (Azad et al., 2023).

The nature of "gray zone" campaigns is: to blur the dividing line between peace and war and between civilian and military endeavours; using civilian instruments to achieve objectives which are in the realm of armed forces; placing all of society at risk and create a sense of ongoing conflict, often without deployment of regular forces and seize a territory. Gray zone embrace of various forms of conflict—terrorism, insurgency and nuclear threats included—to make civilian populations a regular target (Mazaar, 2015). The tools employed are: public diplomacy, clandestine surveillance, uniformed advisors, military exercises. The Army can use exercises to show off specific war-winning capabilities—while hiding the details of how they would be used in a real confrontation (Army Multi-Domain Update: New HQs, Grey Zones, & The Art of The Unfeasible). These methods and tools if employed, should guarantee a successful competition which actually is an essence of today warfare and the multidomain operations (MDO) are design for that purpose. The MDO allows "to orchestrate military activities across all operating domains and environments. Actions can be synchronized with non-military activities and [...] in collaboration with other stakeholders and actors associated with the Alliance. In doing so, this concept will deliver tailored options that build an advantage"[10]—in other words, adjust all the military force and diplomatic, information and economic instruments of power capabilities to the level of violence in the crisis (Mezö, 2022). Flexible use of all assets will allow NATO MDO to conduct high-intensity combat, law enforcement, humanitarian operations and military advisory operations—compete in the ambiguous "gray zone".

Planning and conducting gray zone warfare requires enhancing joint all-domain command and control technology (JADC2), i.e. command, control, communications, computers, cyber intelligence, surveillance and reconnaissance technology and networks, with data security key to promote deterrence. Across land, air, sea, space and cyber domains, JADC2 would enable regular forces to strike faster, more efficiently and

[10] The key differentiating factor between "Joint" and "Multi-Domain" Operations is that "Joint" command structures promote coordination with other armed services, while a "Multi-Domain" mindset must go beyond that to include military and non-military assets, which is. Multi-Domain Operations in NATO—Explained October 5, 2023; https://www.act.nato.int/article/mdo-in-nato-explained/

from farther away (Terry, 2023). JADC2 technology would make MDO more agile if applied in following areas:

- Decentralization of C2: The command posts are decentralized cells in constant motion, connected to each other, but with a low physical and electromagnetic radiation;
- Communication and Information Systems would include existing civilian systems and infrastructure (landlines and mobile phones) and police, rescue or fire department systems. The communication systems should also be supported by autonomous, robotic, stratospheric balloons and airships.
- Manoeuvre: Multi-domain forces in urban environment fight in a dispersed, independent manner, along the general outline set by the operation commander, with the strategic objectives and the desired end result;
- Planning: operational plans should be reviewed and tested with the available tools like: red-teaming, wargames and modelling and simulation (Mezö, 2022).

On 16th September 2021 U.S. Army Europe and Africa activated the 2nd Multi-Domain Task Force at Allen Field on Lucius D. Clay Kaserne. It includes a headquarter, an intelligence, cyberspace, electronic warfare and space detachment and a brigade support company. Its mission is to prevent conflict, shape the security environment and if necessary, conduct large-scale ground combat operations (U.S. Army Europe and Africa "Stronger Together").

Multidomain Forces have the capacity to understand and operate in urban environment, actually they are tailored to the requirements of a dense and complex city fabric shaped by demographic, cultural, economic and political processes. This is true for the following reasons:

1. Cities have multiscale physical structure and operations occur at subterranean, surface and above surface layers.
2. Operation takes a form of competition in various spheres of the city life rather then fight to defeat the adversary. It will target the economics (production and allocation), politics (penetration, participation and legitimacy) and cultural identities in the city (e.g.: insurgency/counterinsurgency operation). The regular forces

attempt to benefit from civil-military cooperation and win support of the local armed forces and law enforcement agencies; economic, financial and political leaders, representatives of local society and refugees, leaders of governmental, non-governmental and international organizations and journalists. Whereas the adversary attempts to neutralize NATO's technological superiority, to conceal their role and interests and to remain on the brink of war. For that purpose, they finance proxy forces, mercenaries, criminals, terrorists, insurgents, extremists, religious fanatics and dissidents who can maintain the fight/competition below the threshold of war to avoid international military aid interventions.

3. Communication technologies: social media, cell net-works and the internet) are a tool to extend the influence and create sympathetic support in formerly neutral forces which could be observed during the Arab Spring protests in Egypt (Wolfel et al., 2021).

The MDO in urban environment are fought in land, air, sea, space and cyber and in the three dimensions: the human dimension, the physical dimension, the information dimension (Rosenberg, 2024), which can be best illustrated in coastal cities, which are significant politically but mainly economically serving as hubs in a global web of markets.

Littoralization may be defined as a "propensity to cluster on coastlines" (Kilcullen) or "“"concentration of economic activity in coastal areas as a result of urban growth, industrial activities, tourism and irrigation"(Zucca and Colombo). The coastal cities enjoy high prosperity since they have a capacity to embrace and enable a sustainable blue economy, attract innovation and new models of tourism, sustainable urban ocean economies for climate adaptation, water security, reducing pollution and fighting a range of other challenges (Economist Impact, 2023). Economic prosperity drive city growth and generates numerous challenges. The dependency between urbanization and income has been proved: the richer the country the more urbanized it becomes. The benefits of urbanization for the city include high density of economic activity, shorter trade links, utilization of human capital, shared infrastructure and division of labour, which results in higher living standards in urban populations than in rural ones. In urban areas there is higher access to higher access to electricity, improved sanitation, drinking water, clean fuels for cooking and heating and child malnutrition are lower.

It should be noted however, that it is difficult to infer causality between urbanization and these examples. Urbanization shows a strong correlation with income and the better access to the above-listed facilities may be merely the effect of higher incomes. There are also considerable inequalities within urban areas—across many low-to-middle-income countries, a high share of the urban population lives in slum households (which lack access to all of the basic resources) (Ritchie et al., 2018). Moreover, there are the decline of some coastal towns is growing problem in the UK. They experience some of the highest unemployment rates and lowest pay. Norfolk and Suffolk, for instance have faced economic recession resulted from the reductions in the shipbuilding industry, fishing and a budget tourism sector. Severe coastal erosion and extreme flooding risk add to the problems. Whereas Newhaven has experienced the decline of traditional manufacturing industries. Flooding risk, air pollution and land contamination issues are additional challenges (Fiorentinoet al., 2024). Both the underdeveloped and well prospering cities with highly developed facilities and infrastructure and the groups of the underprivileged make an area suitable for or prone to military operations (which is also discussed in Chapter 2). Overpopulated cities are culturally, psychologically, economically and geographically valuable for combat as they encompass major communication networks, ground routes, water crossing, logistic and power hubs for sea and air (Anderson, 2023). Littoralization makes added value.

The U.S. *Naval Warfare* defines littoral as follows:

> The littoral is comprised of two segments. The seaward portion is that area from the open ocean to the shore that must be controlled to support operations ashore. The landward portion is the area inland from the shore that can be supported and defended from the sea. (JP 3-32, NDP, p. 2)

Analysing urban littoral environment Milan Vego indicates in detail that the littoral operating areas of both enemy and friendly forces embrace: littoral waters, coasts, offshore islands and the interior within the range of shipborne weapons.

The coastal environment is regarded one of the most challenging and urban littoral combat is "the 'worst of both worlds' and brings together two of the most difficult forms of warfare—urban and amphibious operations" (McCarthy, 2018). Yet, the frequency of contact between opposing forces would be much higher in littorals than on the open ocean and the combat there decisive. The list of weapons used includes: long-range, highly precise anti-ship cruise missiles (ASCMs), land-attack cruise missiles, advanced torpedoes and other smart weapons—would be most likely decisive. It is crucial to point that the characteristics of coastal environment climatic and atmospheric anomalies (changing temperature, pressure, humidity, cloud formation and storm activity), electromagnetic (EM), a large landmass, human-made clutter, the proximity of multiple EM sources and electronic sensors, large number of cellular telephone networks, commercial land-based emitters as television, commercial aircraft and ships may disturb performance of radar, electronic support measures (ESM) and communications systems. Moreover, the surface combat would be fought at close range; encounters would be sudden, short and violent. The intensity of surface and air combat would be very high and consumption of fuel and ammunition would be very huge, consequently logistic support would be detrimental for the outcome (Milan, 2015).

Gunnery Sgt. Neil Lipon from Marine Corps Air-Ground Combat Centre, Calif. (MCAGCC) predicts that:

> A large percentage of the world's population lives in littoral regions. A lot of those littoral regions we'd find ourselves in are urbanized and that's where we may fight in the near future. Marines will be expected to operate in company-sized units in future conflicts. They can be tasked to seize and defend island and coastal territory on their own, while overcoming the friction of operating in urban environments.

The marine corps are trained for so demanding urban operations. On 6th August 2023 at Marine Corps Air-Ground Combat Centre, Calif. (MCAGCC) Range 220. U.S. Marines with 3rd Battalion, 7th Marine Regiment (3/7), 1st Marine Division, conducted Adversary Force Exercise (AFX) as part of Service Level Training Exercise (SLTE)5-23. Marines, Sailors and allied forces trained in urban combat to master urban defence tactics and logistics in a contested environment. Here is the exercise coverage: Marines acted as the Marines of 3/7's opposition to mimic potential future adversaries. The Marines of 3/7 captured urban terrain using amphibious, mechanized tactics while they fought alongside Marines crewing P7 amphibious assault vehicles from 3rd Amphibious Assault Battalion (AAB), 1st Marine Division. After 3/7 seized the terrain, they built defensive positions with sandbags. The Marines of 1st

Combat Engineer Battalion (CEB), 1st Marine Division, embedded with the 3/7 Marines installed concertina wires. Marines trained how to cooperate in a built-up area to minimize casualties and maximize lethality. They also practiced sustainment in a simulated combat environment, preparing for the defence of secured territory and managing logistic support. Food, water, fuel and ammunition allow the troops to sustain and keep pressure on adversaries. Marines from the Republic of Korea (ROK) Marine Corps joined 3/7 in AFX 5-23, to understand and learn each other's tactics and cooperate to maximize effectiveness of deterrence against any opposing forces (Combat Centre News, August 30, 2023).

maritime-focused multi-domain exercise brought together 20 nations, including Sweden, participating for the first time as a full-fledged NATO member. Allied forces engaged in a diverse array of operations at sea, in the air and on land, from complex amphibious manoeuvres to precision air and naval gunnery, anti-submarine warfare and sophisticated mine clearance (BALTOPS 24).

3.4 Within the Society

The experiences from Iraq and Afghanistan have proven that protection of civilians, infrastructure and monuments is of top priority. The joint working group within NATO has been established to monitor the course of military operations in the urbanized areas and formulates recommendations for decision-makers and commanders. NATO should develop technical and organizational aspects of operations as well as professional training of troops. Urban areas will differ in size, shape, demography and functioning (governance, the poor districts, overpopulated districts). Cities will still be economic and political centres, therefore they will be a key element of security and influence the nature of military operations.

In 2016 NATO Defence College organized Urbanisation Seminar Game of a full scale Littoral Urban Operation in a fictitious city of 2035. The analysis of the game outcomes were the foundations for the Capstone Concept—a strategic-level guidance for urban combat out to 2035; it considers city size, climate and development level, covers all domains and NATO's core tasks (NATO Urbanization Project). The project rejects military actions which require cumulation of means and forces, i.e. the tactics of destruction. It promotes the tactics of continuous, sequential military operations based on information and use of precision projection of capacity, measurable effects and adaptation of means to the characteristics of the city. Thus, it may be concluded that the new NATO concept of urban combat focuses on reducing number of casualties in both the adversary and the defender's armies and first of all among civilians. Also, the extent of side-damages will be reduced by precisely combined and selected types of ammunition, forces and subunits.

Cities have always been a destination place for people from rural areas and small, poverty-stricken towns. They have also become attractive for terrorist groups who want to benefit from the city's facilities and capabilities that could enable them to take control over political, economic, socio-cultural and commercial activity since these factors determine the advantage in new wars (Maskan, 2021).

Development of mass media in the 1960s contributed to spread of terrorism which mainly took the form of attacks against military establishments, government buildings, VIPs, or particular racial or cultural groups. In the early 1990s the terrorists shifted towards economic targets to cause economic disruption, social unrest and inflict political pressure on the authorities. The attracts targeted business districts, critical infrastructure (gas, electricity, telecommunication, transportation). Yet, the financial centres became especially endangered due to their clusters of newly designed buildings, their increasingly cosmopolitan communities and first of all the potentially spectacular devastation effects and the significant media attention and publicity the terrorist could receive. The examples of this kind of targeting include: the bombing of the financial districts of New York and Bombay in 1993 and Tokyo, Madrid, Paris, Riyadh and Colombo in 1995. In the UK the main terrorist threat in the 1990s pose the Provisional IRA, which targeted the city of London as the symbol of British imperialism (and state power) and the centre of the British and global financial system.

The 9/11has become a turning point to the perception of terrorism and the subsequent counter-responses (Coaffee, 2009). In response to the terrorist attacks of September 11, 2001, almost all of the twenty major economies (G20) passed anti-terrorism laws that restrict civil liberties and expand the powers of the executive branch to strengthen the national security (Braml, 2021). The military forces, law enforcement bodies and intelligence services developed joint capabilities: created common databases, exchanged personnel, intelligence, technology, expertise and experience and conducted joint training and operations (Gunaratna, 2021).

The methods of attack used by terrorists have evolved. They attack crowded public places using a variety of novel, deployment methods, e.g. a partially successful Vehicle borne improvised explosive device (Vbied) at Glasgow airport in June 2007. The attacks against 'soft targets' have become more likely.

According to Europol prior to the Gaza war, Islamist extremist plots in Europe were decreasing. In 2020 fourteen jihadist plots were recorded in 2021—eleven plots and in 2022 six plots. Whereas in the first nine months since Hamas's 7 October massacre over a dozen such plots have occurred: the 13 October 2023 stabbing at a high school in France, the 16 October 2023 shooting murder of two Swedish football (soccer) fans in Belgium, the 16 October 2023 stabbing in the United Kingdom, the 3 December 2023 killing of a German tourist in France, the 2 March 2024 stabbing of a Jewish man in Switzerland and the 31 May 2024 stabbing attack against a far-right demonstration in Germany. These attacks were conducted by lone actors armed with knives, unlike the large-scale attacks that occurred in Europe in the 2010s. Since the Gaza war began, European security services have baffled apparent plots. The list of the alleged plans includes: to attack Christmas markets and cathedrals in Germany and Austria in December 2023, a concert in Belgium in March 2024, churches and police officers in April 2024. There were also reported an alleged shooting plot under the guidance of IS-KP to attack the Swedish Parliament in March 2024 and an alleged shooting plot against Jewish communities and security officials in the UK in May 2024 (ICTT). The terrorist attacks in Derbent and Makhachkala, two of the largest cities in the Russian Republic of Dagestan neighbouring Chechnya on 23 June 2024 killed at least 21 people and injured 46. The armed men simultaneously attacked two synagogues, two Eastern Orthodox churches and a police post killing 16 policemen and at least 5 other people, including an Orthodox priest. The Islamic State Khorasan (ISIS-K) that claimed responsibility for the Moscow terrorist attack in March 2024, commented on the attacks in Dagestan, that it was carried out by "brothers in the Caucasus who showed that they are still strong". Yet, officially no one has claimed responsibility for the terrorist attacks in Dagestan (UN, 2024).

Moreover, there have also been plots alleged to directly target the Paris Olympics. In April, a 16-year-old was arrested after he announced on social media that he wanted to die a martyr at the Olympics. In May, an 18-year-old was detained under the suspicion of plotting an Islamist-inspired attack at a football stadium during the Olympics (Spaaij &

Zammit, 2024) Just hours before the 2024 Paris Olympic Games opening ceremony France's rail network were hit by arsons defined by authorities as „criminal actions" and a "massive attack to destabilize the railway system" (ABC News, 2024).

Terrorists' attacks very often intermingle with regular crime, violence and gangs' activities leading to the social unrests. Since the end of July 2024 riots broke out at anti-immigration protests in towns and cities across Britain with attacks by far-right groups on hotels, asylum seekers and on mosques. The protests were incited by the incident on July 29 when three girls aged between six and nine were killed, eight other children and two adults were hurt during an attack at a Taylor Swift-themed dance event in Southport, Northern England. Police arrested a 17-year-old male. Disseminated through social media the false information that the suspect was an Islamist migrant led to violent protests. Actually, the teenager has been charged with murder and attempted murder and police informed that he was born in Britain and have not treated the attack as terrorism.

The polarization of society is much deeper and worrying in Sweden which does not have instrument to fight effectively the gangs and "no-go zones", where local clans rule and no one dares to enter without police escort. Cultural, ideological, religious diversity intermingles with contradictory political interests and disinformation:

> The Swedish public debate has already become polarized to a frightening extent, with accusations of racism and fascism polluting the public space. The flames are being fanned from a variety of quarters, from Russian trolls and radical Islamist circles to left-leaning public service media. Surfing on a rising wave of outrage against the government, the opposition has gained a massive lead in the polls. (Hedlund, 2024)

Terrorist attacks pose the serious threat to city-dwellers and public order. The goal of terrorist is not only to target civilians and spread fear, but also to divide city residents. The conflicted communities are more vulnerable to exploitation, radicalization and violent extremism (WEF, 2016). However, analysing the New York City and Washington in terms of terrorist attacks impacts James Harrigan and Philippe Martin provide arguments for the cities being strong enough to resist them and any single terrorists attack will not threaten the economic growth of the city:

> The costs of dealing with terrorism can be viewed as a type of tax that may alter the balance of economic forces that leads to city formation. Although such a change in the attractiveness of doing business in cities may have a very modest effect, or no effect at all, there is at least the possibility that it could lead to a reversal of the circular causation that sustains the existence of cities. (Harrigan & Martin, 2002)

In the face of seemingly intrinsic propensity of the urban areas to become the Gray zone operational environments the question about its resilience capabilities arises. The elaborated framework for building urban resilience focuses at the following activities:

- Respecting IHL and strengthening the protection of civilians in cities at war,
- Strengthening basic services, facilities and infrastructure: healthcare, water, sanitation, electricity and education—to reduce the cumulative impacts of urban warfare and violence.
- Enhancing protection and assistance for the displaced to, between, or within cities,
- Mitigating food insecurity in situations of urban warfare and urban violence also by supporting people's livelihoods and local markets (ICRC 05.08.2022).

The factors accounting for complexity of the urban warfare environment are numerous. They include: size of the armies, urbanization, globalization, technological advances, communication, climate change, population growth, littoralization. These reinforcing trends make city complex operational environment determine to some degree the nature of urban combat, the use of weapons and their results. The attacker has limited cover and concealment.

whereas defenders can hide in any locations, use guerrilla tactics to attack and then disappear back into the urban terrain, canalize attacking forces to ambush sites or down roads filled with improvised explosive devices.

Technology developments offer a wide range of solutions, equipment and weapons: Artificial Intelligence, natural language processing, computer vision, nano- and bio- technologies, advanced big data analytics speed the revolution in sensor technologies, long-range, hypersonic "intelligent" weaponry, "swarms of killing machines" and directed-energy weapons. Yet they all may have reduced effectiveness in densely built-up area.

Operation takes a form of competition in various spheres of the city life rather then fight to defeat the adversary. It will target the economics (production and allocation), politics (penetration, participation and legitimacy) and cultural identities in the city (e.g.: insurgency/counterinsurgency operation). These multidomain operations may prove successful adapting proper set of assets at disposal against the irregular forces which have the defensive advantage in urban terrain.

Taking control over political, economic, socio-cultural and commercial activity of the city give the advantage to terrorist. They achieve it through targeting civilians, spreading fear and dividing city residents. The conflicted communities are more vulnerable to exploitation, radicalization and violent extremism. Building resilience should concentrate on protecting the affected civilians, displaced people and providing basic needs and services.

3.5 Avdiyivka—A Case on Modern Urban Warfare

The war in Ukraine provides many cases of the urban combat. One of them is the long fight for Avdiivka, the key city on the northern-western outskirts of Donetsk. The first defence operation took place in 2014. Since that time the Russians have tried to take over the city several times. Before 2014 Avdiivka with the population of over 30,000 and well operated coking plants was an important economic site. Avdiivka was also an important element of the Ukrainian defence infrastructure for three reasons:

Firstly, the city is located 13 km north from Donetsk, so capturing the city enabled Russians to control the whole area of Donetsk. The fortified city stopped the enemy forces from reaching smaller towns and settlements at the back of the city. To the west of the city the terrain is flat and discouraging for defence operations, thus favourable for operational development of Russian offensive troops at Donetsk. Therefore, all the surrounding neighbourhoods were included in the Ukrainian expanded fortification system. Since the whole region was extensively fortified the Russian troops were not able to bypass it, isolate and surround the defenders. This is the very reason why Avdiivka is important. Secondly, the key highway junction and railway lines in the area are the elements of the infrastructure important for Russians logistic capacities and resources supply. Thirdly, the coking-chemical plant[11] in the city supplied energy raw materials. Pro-Russian separatists attacked Avdiivka in 2014, however they suffered heavy losses and Ukrainians defended their fighting positions. Since 2014 Russians were regularly shelling the city. After the difficult period between 2014 and 2015, the operational-tactical situation lessened although the local actions, fire exchanges and attempts to capture the important objects occurred in many places. In 2017 Russians initiated other attacks at the city mainly at the industrial zone.

The battle of Avdiivka started again in February 2022. When the full-scale war began, Avdiivka became one of the main targets because of its location. On 13th March 2022 the adversary bombed the coking-chemical plant. They did not attack only the energy infrastructure but also city infrastructure causing heavy civil casualties and collateral damage in housing districts. However, Avdiivka was the best fortified site on the front line and therefore Ukrainians managed to counter the next attacks. The front line around Avdiivka ran through the fortified resistance points and defence areas (Wilk, Ber 2023). A strong point of resistance was the coking chemical plant, bombarded and shelled with incendiary ammunition many times. Moreover, the isolated, fortified sectors in the industrial zones were also crucial "centres of gravity".

At the beginning the Russian attacked the city formed in small assault company-based groups in several armoured vehicles and landing troops of 30–40 infantrymen. The next attacks repeated in a wave- pattern were successful for Russians as they took the defenders' fighting positions. The Russians did not enter the city but their success threatened the Ukrainian army and questioned their defence capacity. The Russian tactics also changed. They used small assault infantry groups and sapper units to attack the selected buildings. The assault groups supported by artillery heavily damaged the city structures. The attacks came from several directions simultaneously with massive support of mortars, barrel and rocket artillery. The fierce fight took place in the industrial zone and districts with high buildings. Specially organized automatic grenade launchers teams, anti-tank agents' teams and mortars were used in the direct fight. When fighting in the urbanized environment the infantry troops (predominant in that conditions) used drones to detect defensive fighting positions, observation posts and shell anti-tank guided missiles.

[11] It was a huge workplace with several thousand employees. It was built close to the region where coal—a basic raw material for producing coke—was mined.

Drones were also used to correct artillery fire and coordinate fighting subunits.

The final phase of the battle for the city started on 10th October 2023. The Russians used helicopters, artillery, tanks and armoured vehicles. The Ukrainian army retreated form Avdiivka with a cover of the authorized forces. The cover enabled Ukrainian forces to break away from the adversary and provided a spare time to retreat from the combat zone and avoid encirclement.[12] The last Ukrainian units abandoned their defensive positions in the area of Avdiivka on 17th February 2024.

To sum up, the defence of the city after the Russian aggression continued from 18th April 2022 to 17th February 2024. After conquering Avdiivka, the Russians moved the contact line of the fighting troops further from the Russian supply lines and Donetsk, a key city for the Russian operation. For the Russians Avdiivka located to the north of Donetsk was a convenient area to continue the previously blocked offensive operation. The loss of the city had a huge propaganda value for the Russians. Putin used the information about fighting Ukrainian troops off Donetsk in his presidential campaign (Bailey et al., 2024).

Jarosław Wolski,[13] the journalist covering the war, estimates 53,000–63,000 Russian casualties including 15 000–18 000 killed and 38,000–45,000 injured.[14] Whereas the Ukrainians had 3000–4000 killed and 15 000 injured. Thus, the Ukrainians suffered five times fewer deaths and three times fewer the injured than the Russians. For comparison, ISW experts reported 16 000 to 47 000 killed and injured Russians and the Russians' losses in equipment in 13:1 ratio (Ditrich, 2024). It is estimated that from October 2023 to 23rd February 2024 the Russians lost 690

[12] In the final phase of the battle Avdiivka was defended by 47th Assault Mechanized Brigade, 110 Independent Mechanized Brigade and 3 Assault Brigade "Azov" as a rear guard of.

[13] The publicist and civil analyst of military and defence issues. He has hosted the information channel "Wolski o wojnie" since April 4 2022. Since 2023 he has been an academic teacher in Polish Naval Academy.

[14] The data are approximate since none of the parties reported on losses.

vehicles.[15] The Ukrainians' losses are much smaller and count 75 units of various equipment.[16]

The analysis of the battles for Avdiivka show that cities still play a key role in the modern wars and so in the state defence system. They can restrain the adversary and engage them in gruelling fights reducing their potential. The defensive troops can also break the concentrated forces of the adversary, continue defence for a long time and create favourable conditions for the defenders to take over the initiative at different operational positions.

3.6 Modern Urban Warfare—Discussion and Conclusions

Organization of military operations in urban areas is challenging because of the complex nature of the area itself. The complexity arises from: man-made infrastructure and buildings (industrial, service and sports facilities), topography of the land, population and urban structure. From the military standpoint all these factors interact creating dynamic, complex urbanized areas, yet each city has distinctive physical, political, economic, social and cultural features.

Majority of military instructions and doctrines of most armies strongly recommends especially land forces omitting or isolating urbanized areas. This recommendation however contradicts the intention to capture and control the governmental sites, industrial and business centres, transport and communication nodes. The history of art of war proves that capturing the capital city or political centre may determine the outcome of the campaign. Commanders who lead the operations in the urban area face many limitations:

1. Variety of buildings and objects, especially those inhabited by civilians. The military action is realized with the civilians present in

[15] The Russian losses included 229 tanks, 373 infantry fighting vehicles (IFV) and 16 armoured personnel carriers, 8 rocket launchers, 23 units of artillery, 7 anti-aircraft defence systems.

[16] The Ukrainian losses included 25 tanks, 29 infantry fighting vehicles (IFV), 7 armoured personnel carriers, 1 rocket launcher, 6 units of artillery.

this area.[17] Cities have many functions, some of them concentrate on development of industry others on trade or services. Every urbanized area is unique, differently designed vertically and horizontally (less basements or lack of metro in the suburbs), with different constructions, building materials, streets layout, driveways (increased number of private cars in the suburbs), bypasses, green areas, sports centres and non-built-up areas and the direct surrounding of towns and villages. Original, historical city layouts are untouched for a long time, yet they are developing and expanding to meet new accommodation and transportation needs. Castles, cathedrals and middle-age buildings are located down the narrow, winding streets which mark the old towns (historical centres) of many European cities. In the old town there are few two-way streets. They are too narrow for military vehicles to drive through and to use the mobile weapons. In other cities the centres consist of the high mega-store buildings and commercial networks. Building constructions and materials reflect the functions the city serves, its resources, climate and culture. Chapter one presents the general models of the cities in different regions, which may serve as the base for more detailed military analysis. It is crucial to emphasise that urban ecosystems differ significantly—there are no two identical cities.[18]

2. This variety and uniqueness of city ecosystems create the second limitation. Every single planning and completing the military operations in the urbanized area requires different procedures, military

[17] In October 2004 the British journal „The Lancet"—one of the most prestigious medical journals published the research based on the interviews with civilians of 988 households within the 33 randomly selected administrative districts in the whole Iraq. The researchers estimated an increase in death after the invasion and an additional number of civilians victims which was between 8 and 1,94,000, yet the most probable was 98,000. About half of them died of the violence, most from hand of the US army soldiers.

[18] Districts in Kabul do not have names, are numbered. Every district consists of many blocks which have their names. During ISAF operation; Zabardast, E., Zabardast, S. (2021). *Afganistan. Życie codzienne w Kabulu*. Wydział Informacji o Krajach Pochodzenia. Urząd do spraw cudzoziemców. file:///C:/Users/Lenovo%20E16/Downloads/E_Zabardast,_S_Zabardast_-_Afganistan_%C5%BBycie_codzienne_w_Kabulu.pdf, [accessed: 19.05.2024].

structures,[19] specialized weapons. Contemporary cities cannot function without electricity, food, water supply and waste disposal systems.[20] The life in a city collapses if the access to public transport is limited, the police, fire brigades, hospitals, telephones, media (newspapers, radio, tv) do not work properly.[21] The cases of infrastructure malefactions and their impact are described more extensively in chapter two.

3. The third limitation results from the role of the city in the state defence system. Cities are included in defence planning, in the deterrence strategies because they engage significant capacity of adversary to isolate them and then capture particular neighbourhoods, settlements and districts. In the Eastern Europe the villages, towns and cities are loosely scattered about several kilometres far from one another. Thus, urban operations are long and devastating when the offensive forces try to break the defendants' will by bombing and artillery shelling as the history confirms.[22] Moreover, urbanization influences not only the nature of military operations but also peacebuilding and humanitarian operations (the issue discussed in chapter three). It may happen that while the military action is continued in one part of the city the humanitarian operation is realized in another part.

In such operational conditions decision-makers and commanders have three options to choose from: The first option is to clear the city.

[19] Assault groups, combat team, special reserve, clearing units.

[20] In March 2023 4.5 thousand tons of waste were lying in the streets as the mass protests against the retirement system reform disturbed garbage collecting. Wastes piled around such tourists' attractions like Eiffel Tower and Arc de Triomphe.

[21] In September 2012 in Madrid and Barcelona public transport services workers went on strike. In the two biggest Spanish cities busses, metro and trains were running five times less often than usually. The reason for the strike were reduction of wages and a plan to divide the company into four smaller ones.

[22] Bombing of Dubrovnik can be an example. The siege of the city started on 1st October 1991. In spite of the fact that Dubrovnik had already been protected under the international law on protection of ancient buildings the Yugoslavian National Army was blocking and shelling the city from land, sea and air since October to December 1991 targeting mainly civilians and civil objects. As the result almost 70% of the historic buildings was damaged. The renovation cost of the old city amounted to \$ 10 m. As the city was on the UNESCO heritage sites list it quickly gained the funds for the renovation so at present the city is completely restored with almost no traces of war.

The second one is the situation when the troops cannot clear the city because of its topography, political or military reasons—the city must be conquered. The third option is a siege of the city, i.e. isolating it from the main forces. The city engages much of the adversary capacity but also suffers losses from the airstrikes and artillery shelling. The siege of the city is a prolonged military blockade of food supply and military support for defenders. These actions last till the food, water, ammunition and medical aid are depleted.[23] Moreover, the prolonged siege may break the defenders' morale.[24] On the other hand, city defence as a symbol of resistance against the adversary may strengthen national pride and social resistance. In present times the symbols of city resistance are: Dubrovnik,1991–1992, Basra, 2008.

References

ABC News France Train Lines Hit By Arson Attacks Just Hours Before. (2024). Paris Olympics Opening Ceremony.

Allen, B. (2024). *Top Ten Emerging Technologies for DOD and National Security. Velocity Magazine*. Special Edition Report.

Amnesty International. (2021, March 3). *States Should Curb Urban Slaughter of Civilians from Explosive Weapons*, https://www.amnesty.org/en/latest/press-release/2021/03/states-should-curb-urban-slaughter-of-civilians-from-explosive-weapons-2/. July 31, 2024.

Anderson, M. (2023, May–June). Decided Among Cities. *Military Review*, 25–32.

Azad, T. M., Haider, M. W., & Sadiq, M. (2023). Understanding Gray Zone Warfare from Multiple Perspectives. *World Affairs, 186*(1), 81–104. https://doi.org/10.1177/00438200221141101

Bagsaw, S. (2023, December). The 2022 Political Declaration on the Use of Explosive Weapons in Populated Areas: Tool for Protecting the Environment in Armed Conflict? *International Review of the Red Cross*, IRRC No. 924. https://international-review.icrc.org/articles/the-2022-political-declaration-on-the-use-of-explosive-weapons-in-populated-areas-924. July 31, 2024.

[23] For example—the city lies at the river or in the valley or is a political centre.

[24] For example—during WWII Hitler allowed Albert Kesselring to leave Rome without fighting on 3 June 1944 mainly because it was an ancient city and a capital of his Italian ally- Benito Mussolini. Moreover, Vatican was situated within the city, which is a place of worship for Bavarian Catholics. The historical artefacts in Vatican were unique. When the allies arrived they reported that all key bridges on the Tiber and historical buildings were untouched.

Bailey, R., Evans, A., Wolkov, N., Mappes, G., & Kagan, F. W. (2024, April 28). *The Battle for Avdiivka Enters its Third Month: Day 656 of the War*. Institute for Study of War. https://www.understandingwar.org/backgrounder/russian-offensive-campaign-assessment-april-28-2024.

BALTOPS 24 Concluded; Demonstrates NATO's Unwavering Commitment to Baltic Sea Security, June 24, 2024. ACNato, https://ac.nato.int/archive/2024/BALTOP24-concluded

Betz, D., & Stanford-Tuck, H. (2019, October). The City Is Neutral: On Urban Warfare in the 21st Century. *Texas National Security Review, 2*(4), 60–87.

Bosio, N. (2020, March 5). *What Is War? Defining War, Conflict and Competition*. Australian Army Research Centre. https://researchcentre.army.gov.au/library/land-power-forum/what-war-defining-war-conflict-and-competition

Braml, J. (2021). *Peace and Security. Anti-Terrorism Laws nd Powers*. An Inventory of the G20 States 20 Years after 9/11. Friedrich Ebert Stiftung.

Braml, J., & Lambrecht, C. H. (2021, September). *Antiterrorism Laws and Powers,* Friedrich-Ebert-Stiftung, https://ny.fes.de/article/anti-terrorism-laws-20-years-after-9-11.html

Coaffee, J. (2009). *Terrorism, Risk and the Global City Towards Urban resilience*. Ashgate Publishing limited.

Combat Center News, Twentynine Palms, California.

Danielsson, A. (2025). The Emergence of a Military Urban in and of War. *Annals of the American Association of Geographers, 115*(2), 404–418. https://doi.org/10.1080/24694452.2024.2422856

Depczyński, M., & Elak, L. (2020). *Rosyjska Sztuka Operacyjna. Teoria i praktyka. Fundacja Historia i Kultura*, 301–305.

Ditrich, R. (2024, February 24). *Awdijiwka padła. Już 10 lat temu Rosjanie próbowali zdobyć to miasto*. Forsal.pl. https://forsal.pl/swiat/bezpieczenstwo/artykuly/9441219,awdijiwka-padla-juz-10-lat-temu-rosjanie-probowali-zdobyc-to-miasto.html

Economist Impact. (2023, April 17). *Blue Cities: How the Ocean Can Be an Economic Engine for Sustainable Urban Development*. https://impact.economist.com/ocean/sustainable-ocean-economy/blue-cities-how-the-ocean-can-be-an-economic-engine-for-sustainable-urban. August 17, 2024.

Conflict in Syria. (2017). Emerging Military Technologies Applied to Urban Warfare, *International Review of the Red Cross, 99*(3), 1161–1174. https://doi.org/10.1017/S1816383118000395

European Security and Defence. (2022, March 8). *Can Technology Transform Future Urban Combat?* https://euro-sd.com/2022/03/articles/exclusive/25293/can-technology-transform-future-urban-combat/

Fiorentino, S., Sielker, F., & Tomaney, J. (2024, March). Coastal Towns as 'Left-Behind Places': Economy, Environment and Planning. *Cambridge Journal of*

Regions, Economy and Society, 17 (1), 103–116, https://doi.org/10.1093/cjres/rsad045

FM 90-10-1.HEADQUARTERS DEPARTMENT OF THE ARMY Washington, DC, 30 September 1982.

Freedberg, Jr., S. J. (2018, December 07). *Army Multi-Domain Update: New HQs, Grey Zones, & The Art of The Unfeasible*. https://breakingdefense.com/2018/12/army-multi-domain-update-new-hqs-grey-zones-the-art-of-the-unfeasible/

Garstka, J. (2003). Walki w miastach. Obrona w terenie zurbanizowanym -trochę teorii. *Raport: Wojsko-Technika-Obronność*, nr 4, 24–26.

Georgia Tech. (2022, March 21). *How Social Media is Changing the Way Citizens and Companies React to War*. https://www.scheller.gatech.edu/news/2022/how-social-media-is-changing-the-way-citizens-and-companies-react-to-war.html

Gormezano, D. (2024, April 8). *In Ukraine's Donbas, Ten Years of War and Russification*. France24. https://www.france24.com/en/europe/20240408-ukraine-donbas-ten-years-of-war-russification-russia-donetsk-luhansk

Gunaratna, R. (2021, September 1). *The 9/11 Effect and the Transformation of Global Security*, https://www.cfr.org/councilofcouncils/global-memos/911-effect-and-transformation-global-security. August 13, 2024.

Halloran, T. (2022, March 21). *How Social Media is Changing the Way Citizens and Companies React to War*.Georgia Tech. https://www.scheller.gatech.edu/news/2022/how-social-media-is-changing-the-way-citizens-and-companies-react-to-war.html

Harrigan, J., & Martin, P. H. (2002, November). Terrorism and the Resilience of Cities. *Economic Policy Review, 8*(2).

Hedlund, S. (2024, February 9). *Sweden Looks into the Abyss, GIS Report*. https://www.gisreportsonline.com/r/sweden-immigrants-crisis/

Hills, A. (2004). Future War in Cities. Routledge.

ICRC (2022, August 5). *The ICRC at the World Urban Forum: Cities' resilience is key in armed conflict and violence*, https://www.icrc.org/en/document/icrc-world-urban-forum-2022. August 17, 2024.

Jagiellonian University in Krakow. (2023). *Locating an ancient battleground*. https://en.uj.edu.pl/en_US/ju-research/-/journal_content/56_INSTANCE_2XezEHy2NT5h/81541894/141050421

Joint Publication (JP) 3-32, Command and Control of Joint Maritime Operations, 8 June 2018, and JP 2-01.3, Joint Intelligence; Naval Warfare. Naval Doctrine Publication 1, April 2020.

Jordan, J. (2020). International Competition Below the Threshold of War: Toward a Theory of Gray Zone Conflict. *Journal of Strategic Security,* 14(1), 1–24. https://www.jstor.org/stable/26999974

Kaldor, M. (2014). *Human Security: Political Authority in a Global Era*. In. Martin, M. & Owen, T. (Eds.). Routledge Handbook of Human Security.

King, A. (2021). *Urban Warfare in the Twenty-First Century*. Polity Press.

Krawczuk, A. (1990). *Gajusz Juliusz Cezar*. Ossolineum

Krawczyk, R. (2012). *Podbój Kanaanu w świetle Księgi Jozuego*. Historia i wiara. Studia Warmińskie 49 (2012) ISSN 0137–6624

Laub, Z. (2023, February 14). *Syria's Civil War: The Descent Into Horror*. Council on Foreign Relations. https://www.cfr.org/article/syrias-civil-war

Makowiec. P. & Mroszczyk,M. (2012). *Taktyka walki w terenie zurbanizowanym*. Difin.

Marine Air Ground Task Force Training Command and Marine Corps Air Ground Combat Centre; 30 August 2023 | Cpl. Jonathan Willcox Marine Corps Air Ground Combat Centre Twentynine Palms; https://www.29palms.marines.mil/Articles/Article/3511053/37-prepares-to-fight-sustain-defend-urban-terrain-in-the-conflicts-of-tomorrow/. August 8, 2024.

Maskan, I. (2021, November). *Urbanization of Conflict*. Globalization and Increased Urban Warfare, https://www.dlpforum.org/2021/11/30/globalization-and-increased-urban-warfare/

Mazarr, M. J. (2015*). Astering the Gray Zone: Nderstanding Gray Zone Conflict*. Strategic Studies Institute, US Army War College. https://www.jstor.org/stable/resrep12018.9

McCarthy, D. (2018, April). The Worst of Both Worlds: An analysis of urban littoral combat. *Australian Army Occasional Paper Series*. Australian Army Research Centre. https://researchcentre.army.gov.au/library/occasional-papers/worst-both-worlds-analysis-urban-littoral-combat

Mezö, A. (2022). Multi-Domain Operations in An Urban Environment. *The Three Swords*, 38.

Milan, V. (2015). On Littoral Warfare. *Naval War College Review, 68*(2), Article 4. Available at https://digital-commons.usnwc.edu/nwc-review/vol68/iss2/4

Multi-Domain Operations in NATO—Explained. (2023, October 5). https://www.act.nato.int/article/mdo-in-nato-explained/

Odziemkowski, J. (1989). *Warszawa w wojnie obronnej 1939*. PWN.

Preparation of the Operational Environment 21 May 2014.

Ritchie, H., Samborska, V., & Roser, M. (2018/2024 rev.). *Urbanization. Our World in Data*. https://ourworldindata.org/urbanization

Rosenberg, B. (2024, May 23). *How the Army is Driving Enterprise Training Across Five Warfighting Domains and Three Dimensions*, https://breakingdefense.com/2024/05/how-the-army-is-driving-enterprise-training-across-five-warfighting-domains-and-three-dimensions/. August 16, 2024.

Sapaaij, R., & Zammit, A. (2024, June 27). *The Terrorism Threat to the 2024 Paris Olympics: Learning from the Past to Understand the*

Present. ICCT. https://www.icct.nl/publication/terrorism-threat-2024-paris-olympics-learning-past-understand-present

Spenser, J. (2021, January 12). *The Eight Rules of Urban Warfare and Why We Must Work to Change Them*, https://mwi.westpoint.edu/the-eight-rules-of-urban-warfare-and-why-we-must-work-to-change-them/

STO-MP-MSG-207. Smith, M., & Robson, D. (2023). *A Concept for the Modelling and Simulation of Complex Urban Environments*. NATO DSTL/CP152445.

Terry, J. (2023, March 31). Gray Zone' Competition—The Race for Multi-Domain Capability. *National Défense.* https://www.nationaldefensemagazine.org/articles/2023/3/31/industry-perspective-gray-zone-competition-the-race-for-multi-domain-capability

The Head of France's Rail Network Called It a "Coordinated Attack." Jon Haworth July 26, 2024. https://abcnews.go.com/International/france-train-lines-hit-arson-attacks-hours-2024/story?id=112296820

The NATO Urban Warfare Project. (2017, Jan. 25). https://mwi.westpoint.edu/event/nato-urban-warfare-project/, https://www.jwc.nato.int/images/stories/_news_items_/2019/three-swords/NATOUrbanization_2035.pdf

The Political Declaration on Strengthening the Protection of Civilians from the Humanitarian Consequences arising from the use of Explosive Weapons in Populated Areas Section 3, 3.1., 3.3. (2025). UNODA.

The Terrorism Threat to the 2024 Paris Olympics: Learning from the Past to Understand the Present 27 June 2024, https://www.icct.nl/publication/terrorism-threat-2024-paris-olympics-learning-past-understand-present

U.S. Army Europe and Africa "Stronger Together". (2021, Sept. 8). https://www.europeafrica.army.mil/ArticleViewPressRelease/Article/2767270/media-advisory-us-army-europe-and-africa-to-activate-2nd-multi-domain-task-force/

UN. (2024, June 27). *Russia: UN Experts Condemn Brazen Terrorist Attacks in Dagestan*, https://www.ohchr.org/en/press-releases/2024/06/russia-un-experts-condemn-brazen-terrorist-attacks-dagestan

Urban Combat Advanced Training Technology. Final Report of Task Group 032, AC/323(MSG-032)TP/293, (2010). Brussel.

Weissmann, M. (2023). Urban Warfare: Challenges of Military Operations on Tomorrow's Battlefield. In Weissmann, M., & Nilsson, N. (Eds.), *Advanced Land Warfare: Tactics and Operations.* Oxford Academic, https://doi.org/10.1093/oso/9780192857422.003.0007. Accessed October 17, 2024.

Wilk, A., & Ber, J. (2023, December 12). *The Battle for Avdiivka Enters Its Third Month: Day 656 of the War.* Centres for Eastern Studies. https://www.osw.waw.pl/en/publikacje/analyses/2023-12-12/battle-avdiivka-enters-its-third-month-day-656-war

Wolfel, R. L., Richmond, A., & Ridgeway, J. (2021, January–February). Dense Urban Environments The Crucible of Multi-Domain Operations. *Military Review*, 22–32.

World Economic Forum. (2016, December13). *Terrorists Want to Destroy Our Cities: We Can't Let Them*, https://www.weforum.org/agenda/2016/12/terrorists-want-to-destroy-our-cities-we-can-t-let-them/

Zucca C., Biancalani, R., Hamrouni, H., Attia, R., & Bunning, S. (2009). *Guidelines for the identification, selection and description of nationally based indicators of land degradation and improvement.* https://openknowledge.fao.org/server/api/core/bitstreams/70c271b4-be76-444f-8b9a-568d7cc9c4e2/content

CHAPTER 4

Disaster Risk Management for Future Crises and Conflicts

Abstract The shift of DRM concepts from solely scientific and technical towards socially sensitive has resulted from disasters that had occurred, their nature and impact they had on societies. The present DRM strategies are still evolving to recognize unexpected events resulting from climate change and protracted crisis and interaction with society and focus on adaptive capacity, transformability, learning, and innovation. These approaches engage various stakeholders and civilians/local communities into the DRM processes embracing technical and human dimensions: infrastructure protection and social protection. In practice, the DRM conceptualized this way becomes an integrated structure in which collective strength creates the added value. The case study of the flood in the southern-western Poland (September 2024) illustrates how DRM system integrates all necessary entities, especially military units to respond to the crisis effectively.

Keywords The integrated disaster risk management · Civil-military cooperation · Flood · Socially sensitive DRM

S. Kowalkowski et al., *Cities in Crisis*,
https://doi.org/10.1007/978-3-032-08843-7_4

4.1 Evolution of Perspectives

The discussion on resilient cities has been commenced in chapter two and focuses on urban environment, infrastructure and services which make the city more adaptable and accessible for the citizens affected by variety of shocks (overpopulation, economic stagnation, natural disaster). At that point, it is crucial to direct an attention to the cities facing the compound and protracted crises the impact of which is intensified by ongoing war. Military conflicts occur alongside lengthy food crises, insufficient governance and institutional capacity making the population acutely vulnerable to death, disease and disruptions in livelihoods over a prolonged period of time. Services essential for the civilian population during conflict are interdependent - the disruption of one may result in multiple services being disrupted or collapsing. Electricity supply ensures the delivery of water and sanitation, solid waste disposal and the cold chain (i.e. the storage, management and transportation in specific temperatures of essential items such as foodstuffs and medicine). A reliable supply of safe water, sanitation and electricity is indispensable for hospitals, food production and distribution. Lack of the basic services (water and sanitation, food, heating and healthcare) leads to disease and death and causes mass displacement especially from regions with freshwater shortages or impacted by extreme weather events (e.g. floods, heat waves, wildfires) and climate change. Women and children, older people and persons with disabilities are the most affected. Their suffering is exacerbated (Giorgou & Zeith, 2023) and therefore DRM strategies need to be shifted from protecting the infrastructure itself to protecting infrastructure in order to protect and save people and eventually build resilient communities. United Nation Disaster Risk Reduction history demonstrates this evolution path.

In 1970s, it was observed that the consequences of natural hazards were becoming so excessive and severe that the emphasis on pre-disaster planning and prevention became priority. The methodology for risk and vulnerability analysis was elaborated. In 1980s.

the International Framework of Action for the International Decade for Natural Disaster Reduction (IDNDR) was developed. Its intention was to reduce, through concerted international action, especially in developing countries, loss of life, poverty damage and social and economic disruption caused by natural disasters. In 1990s a crucial moment in IDNDR was the adoption (in 1994) of the Yokohama Strategy for a Safer World: Guidelines for Natural Disaster Prevention, Preparedness and Mitigation, containing the Principles, the Strategy and the Plan of Action (Yokohama Strategy) at the World Conference on Natural Disaster Reduction. In 1994 The Yokohama Strategy marked the beginning of a significant shift in disaster reduction approaches. While IDNDR was largely scientific and technical strategy, the Yokohama Strategy emphasized the importance of socioeconomic factors and the role of human actions in reducing the vulnerability of societies to natural hazards and disasters. In 2000s the IDNDR was succeeded by the International Strategy for Disaster Reduction (ISDR) which would enable communities to build their resilience to the effects of natural hazards and related technological and environmental disasters, thus reducing the compound risk posed to social and economic vulnerabilities of societies. The Yokohama Strategy contributed greater official and public understanding of the impact of disasters on the economic, social and political fabric of societies.

This shift of DRM concepts from solely scientific and technical towards socially sensitive resulted from disasters that had occurred, their nature and impact they had on societies. The times of World Trade Centre Terrorist Attack 1993 were dominated by the DRM strategies focused on physical infrastructure protection. The resilience was understood as the robustness and redundancy of systems, networks or elements and the capability to maintain equilibrium. As the hazards have become more complex with time of the terror attacks in New York, Washington 2001; Madrid 2004 and London 2005 and natural disasters such as Northeastern blackout in the US and Canada 2003, Hurricane Katrina in 2005 or the UK flooding in 2007 along the increasing complexity of more and more technologically advanced and integrated systems and increased dependency on natural resources consumption, the DRM systems required other more comprehensive, all hazard strategy to address social issues as well. The essence of this approach is included in the Sendai Framework for Disaster Risk Reduction 2015–2030 (Sendai Framework) adopted in 2015.

> The key lesson is that disasters are social constructs. People are vulnerable to the impacts of climate change and natural hazards due not just to their geographical context, but their financial, their social status, their cultural status, their gender status, their access to services, their level of poverty, their access to decision making and their access to justice. Good development is good adaptation and its good risk reduction. (Arnold, 2012)

These vulnerabilities may lead to unexpected decisions, behaviours and policies creating highly unexpected situations the citizens operate in:

> [...]widely divergent forms of disaster, under the right historic circumstances and given the right response, can trigger larger-scale crises and usher in a highly imbalanced form of urbanization (Greenberg, 2017)

The DRM strategies are still evolving to recognize unexpected events resulting from climate change and protracted crisis and interaction with society to focus on adaptive capacity, transformability, learning and innovation. These approaches are sensitive to society in double manner by building the mechanisms of engaging various stakeholders and civilians/ local communities into the DRM processes and by embracing technical and human dimensions: infrastructure protection and social protection. In practice, the DRM conceptualized this way may become an integrated structure in which collective strength creates the added value.

4.2 The Integrated Disaster Risk Management

Sandoval et al. propose Disaster Risk Management model integrated from four interconnected dimensions: sectoral, spatial/hierarchical, temporal and externally with other cross-cutting societal challenges. The sectoral dimension integrates different kinds of actors, either within (internally) or among them (externally) such as classes, groups, institutions and social sectors: Private, Public, Academia, Civil society, Tourism, Critical infrastructure. The special dimension integrates DRM system at different geographical scales or levels. Actors or mechanisms of DRM operate from the national to the local level simultaneously and hierarchically. At this scale, states are also integrated into international and/or upper-regional levels, such as the EU Civil Protection Mechanism in Europe. The third, the temporal dimension integrates actors or mechanism of DRM that work throughout the main DRM phases (prevention, preparedness, response and recovery,), with special focus on prevention and transformative resilience. Fourth dimension integrates societal processes, such as climate change, development and urbanization. This dimension type of integration means that DRM policies/procedures are embedded in systemic operations of other societal processes and that it goes beyond the typical civil protection or risk management organizations (Sandoval et al., 2023).

Integration determines how effectively the goals and task are realized; however, it cannot be achieved by one-time action, it is a process (Alcántara-Ayala et al., 2022).

> Organizational integration is more than coordination or restructuring. It is a deliberate and fundamental shift in how people think about their work, their communities and their partners. Integration focuses on outcomes and results, rather than activities and processes. It optimizes human and financial resources and helps prioritize partners, projects and target markets. (Bischoff, 2021, p. 2)

The integration may be accomplished through (Goh, 2024):

1. Cross-Functional Collaboration contributes to diverse expertise and more effective response. It results in:

> the implementation of integrated and inclusive economic, structural, legal, social, health, cultural, educational, environmental, technological, political and institutional measures that prevent and reduce hazard exposure and vulnerability to disaster, increase preparedness for response and recovery and thus strengthen resilience (Sendai Framework, p.12)

This line of action requires "mov[ing] from traditional multidisciplinary (i.e. working within disciplinary silos) or interdisciplinary approaches (i.e. working across disciplinary silos but using disciplinary methods) of risk assessment to integrated transdisciplinary approaches" (Science for DRR, 2020, p. 57). However, there are some challenges to face on the way: a lack of common understanding of the concept and nature of risk, limited awareness of the contributions different disciplines can make to risk assessments, the willingness and capacity of stakeholders to engage with opposing views, power relations or the choice of the 'right' stakeholders.

2. Unified communication is a critical tool in emergency preparedness and response and requires consistent internal and external communication protocols including designated spokespersons, channels (traditional media, social media) and messaging strategies and tools (e.g. digital or analogue risk maps). Well organized risk communication enables the real-time exchange of information, advice and opinions between experts, officials and people who face a threat. It

helps people at risk to make informed decisions and behave in a safe manner.

3. Comprehensive Risk Assessment

Risk assessments allow to understand possible risks and their drivers, dynamics and potential consequences, identify vulnerabilities and prepare for a spectrum of scenarios. Methods employed for this task differ and depend on context and the complexity of the types of risks. They include: qualitative, semi-quantitative and quantitative methods.

4. Strategic Partnerships

Collaboration among different groups and sectors of society at local, national, regional and global levels enhances resource-sharing, mutual support and collective capabilities in times of crisis. Trust and long-term partnerships are necessary to overcome the differences in responsibilities, interests, culture and experiences and to build the integrated disaster risk management systems within and across all sectors. The partnership at regional and global level is especially desired to address the needs and vulnerabilities of disaster-prone countries, the least developed countries, small island developing States, landlocked developing countries and African countries. The Sendai Framework reads: "It is urgent and critical to anticipate, plan for and reduce disaster risk in order to more effectively protect persons, communities and countries, their livelihoods, health, cultural heritage, socioeconomic assets and ecosystems and thus strengthen their resilience" (p. 10). This kind of collaboration is executed by Directorate General of European Civil Protection and Humanitarian Aid, The Euro-Atlantic Disaster Response Coordination Centre (EADRCC) or United Nation Disaster Risk Reduction Office.

5. Training Programs

To promote regular and scenario-based disaster preparedness, response and recovery exercises: evacuation drills, training to ensure effective response to disasters and related displacement, access to safe shelter, food and non-food relief supplies.

6. Continuous Improvement Culture

Cultivating a culture of continuous improvement means learning from the previous crisis experience, refining strategies and standards, promoting culture of prevention by dissemination of disaster risk information, providing education and training on disaster risk reduction. (Sendai, p. 26).

7. Technology Integration

Implement innovative technologies for crisis communication, response and recovery. They include multisectoral forecasting and early warning systems, social technologies and hazard-monitoring telecommunications systems, cybersecurity measures (Table 4.1).

Table 4.1 Differences in disaster impact between rich and underdeveloped countries

Differences in disaster impact between developed and underdeveloped countries	
Developed countries	*Underdeveloped countries*
Tend to suffer higher economic losses, but have mechanisms in place to absorb these costs	Have less at risk in terms of financial value, but maintain little or no buffer to absorb even low financial impacts Economic impact can be significant and social development ultimately inhibited
Employ mechanisms that reduce loss of life, such as early warning systems, advanced technologies, enforced building codes and zoning	Lack the resources necessary to take advantage of advanced technologies (building codes and zoning) even if these solutions exist
Have immediate emergency and medical care systems that increase survivability and contain the spread of disease	Sustain massive primary and secondary casualties
Transfer much of personal, private and public risk to insurance and reinsurance providers	Generally, do not have well established insurance systems. Transfer funds from development programs to emergency relief and recovery

Source Based on: Coppola, D. P. (2015). The management of disasters. In *Introduction to International Disaster Management* (Third Edition). Science Direct, p. 21, 22. https://booksite.elsevier.com/samplechapters/9780750679824/02~Chapter_1.pdf,

Method of analysis: the development of flood in Poland in September 2024 and the actions of the DRM system are presented through the response and recovery phases with conclusions on impact and lesson learned. The integration of DRM system is assessed by the following indicators: cross-functional Collaboration, communication, technology integration, culture of improvement, training programs, strategic partnerships, comprehensive risk assessment.

4.3 Flood in Central Europe September 2024 – A Case on Civil Protection in Poland

In the beginnings of September 2024 Storm Boris and extremely humid Genoa low brought heavy rainfall and caused flood in Austria, the Czech Republic, Poland, Romania and Slovakia and then onwards Germany, Hungary and Italy killing at least 21 people and forcing evacuations.

In Poland the state and local authorities monitored the situation, issued flood warnings and prepared locals to fight the oncoming flood waves. Volunteers and emergency workers secured river banks of the Oder Basin, in Wroclaw, Lower Silesia, in various parts of the Silesian and the Opole provinces (Fig. 4.1) and advised residents to evacuate (Stezycki et al., 2024).

On 15th September the Racibórz Dolny dry reservoir on the Oder River was activated to mitigate the flood risk (TVP World, 2024, Sep. 15). This biggest hydrotechnical construction in Poland is to protect the Lower and Upper Silesia and Opole territories from flooding. It measures 26 km kw. and has a capacity of 185 m cubic meters. Its dams are over 20 km long. On 17th September at 12 o'clock the reservoir was filled with water in 80 per cent and the next day 18th Sept. at 8 o'clock the water level lowered to 74.5 per cent of its capacity (Business Insider, 2024b, Sep. 19).

4.3.1 Response

As the situation was worsening Prime Minister Donald Tusk announced a state of natural disaster in flooded areas (Fig. 4.1) under The Act Of 18 April 2002 on The State of Natural Disaster which enables conducting

Fig. 4.1 High alerts and water levels on Polish rivers from 10 to 17th September 2024. (*Source* Instytut Meteorologii i Gospodarki Wodnej—Państwowy Instytut Badawczy Tygodniowy Biuletyn Hydrologiczny (2024, Sepr. 10–17); IMGW-PIB https://res2.imgw.pl/products/hydro/monitor-lite-products/BIULHYDTYGPDF.PDF?ts=1726747200000)

evacuation, rescue operations and disposal of financial support for the victims (EuroNews, 2024a, Sep. 17).

Under this Act the government may also introduce price controls to manage the affordability of basic goods and services on the market. Donald Tusk at the DRM team meeting in Wrocław was informed that the prices in the flooded Kłodzko are significantly higher and he warned all those who wanted to make money on flood and human tragedy that he would not hesitate to use this legal tool (Business Insider, 2024a, Sep.18).

Military support was offered by German and Turkey soldiers while American troops were already helping flood-affected communities in southern Poland in accordance with The Act of 23 September 1999 on the rules of stay of foreign troops on Polish territory and the rules for their movement within that territory (Starcevic, 2024).

The above-mentioned Act of 18th April 2002 on the State of Natural Disaster, The Act of 26 April 2007 on Crisis Management, the Act of 11 March 2022, on the defence of the Fatherland regulate the deployment of Polish military units and their support in crisis situations. These regulations were embodied this time since The European Union's authorities agreed that the floods in Central Europe and the deadly wildfires in Portugal are proof of a "climate breakdown" (EuroNews, 2024b, Sept. 19) that must be dealt with urgently.

Over 14 thousand soldiers are engaged in evacuation and rescue operations in the flood-affected areas. They distribute food and water and evacuate people from hard-to-reach areas with helicopters and monitor the threatened areas and collect information on the evolving crisis with drones.

Military Field Hospital created the medical point in Nysa after the local hospital was flooded and 144 patients evacuated. The citizens have already received medical treatment. Emergency unit accepts 120 people a day. The hospital has all necessary medicaments to react in emergency including tetanus toxoid (Pakuła, 2024). I Military Field Hospital in Bydgoszcz is ready to provide emergency support. The Deputy Prime Minister Wladyslaw Kosiniak-Kamysz appreciated cross – functional cooperation of health care system, soldiers and civilians at a press conference in Nysa:

> A field military hospital has been developed in Nysa, with the participation of the management and doctors, nurses of the Nysa hospital, which had to be evacuated. The field hospital in Nysa is in operation. If someone needs help, there is this initial help, but also that of a specialized nature. If necessary, a helicopter is available and transport by air to other hospitals [...] After a night operation to strengthen the embankments with helicopters, Nysa defended itself. It was a great commitment of all forces and resources. We thank all the residents, all the people of goodwill who are helping today and rescued then. Thank you to all the volunteers, the services and all those who are in action all the time. Thank you very much to the doctors and nurses, because we mention many circles, but medicine and the military and civilian is mobilized, involved. If someone needs help, it is this initial

help, but also of a specialized nature, of course, wound supplies. There are already the first admitted patients... .(Medexpress, 2024)

In Lądek-Zdrój, Stronie Śląskie and Kłodzko container kitchens were installed to distribute warm meals to the victims. The most affected are provided also with the dry food ratios.

The citizens who did not have an access to water was supported by soldiers who distributed drinking water with 48 water tankers.

The whole rescue operation is commanded and coordinated by commander of Territorial Defence Forces who cooperates also with all emergency services. The Military units mainly concentrated on:

- Strengthening flood embankments in most threatened areas: Krosno Odrzańskie, Szprotawa, Oława, Nysa, Brzeziny, Kędzierzyn-Koźle and Jelcz Laskowice to lessen the impact of flooding. Evacuating citizens from Lewin Brzeski, Grodkowo, Nysa and Prudnik. Military units together with firebrigades evacuated over 242 people including 7 children and animals (circa 4000 people);
- Continuing preventive and protective measures to safeguard the property of the evacuated. Military Police together with Police patrol the threatened areas;
- Removing flood damage and gradually restoring transport infrastructure;
- Disinfecting buildings and equipment to prevent spread of infections and diseases (Celej, 2024).

4.3.2 Disinformation

The crisis situation became a fertile ground for disinformation. Internal Security Agency confirms the information about the man disguised as a soldier who threatens locals with blowing up the flood embankments. He is driving the car with fake registration plates (Brzozowski & Rybak, 2024). Moreover, the fake news with links is disseminated via SMS as the Alert RCB (alerts from the Government Security Center). The authentic alerts do not have links with an instruction to click on.

To control the situation the ministers and other entities collaborated while informing public opinion. Donald Tusk and minister of National Defence Władysław Kosiniak-Kamysz, minister of Internal Affairs Tomasz

Siemoniak and Minister of infrastructure Dariusz Klimczak stayed in the affected region. Every day they organized the briefings at which representatives of local government, Fire services, Police, Polish Waters, Institute of meteorology and water management reported on the evolving crisis. This strategy enhanced the quality of communication and information content as well as comprehensive risk assessment.

Whereas new technologies like dedicated satellite monitoring enhance the situational awareness.[1] On Friday, September 13, 2024 the Crisis Information Centre of Space Research of the Polish Academy of Sciences (CIK CBK PAN) launched a dedicated satellite monitoring system. The reason for the earlier start was the flood threatening the southern-western Poland (its launch was planned in spring 2025). This is the project of the European Space Agency (ESA) aimed at increasing the crisis response capabilities.

CIK CBK PAN in cooperation with the Polish-Finish company ICEYE. ICEYE operates the world's largest civilian constellation of radar satellites, providing critical data on flooded areas to support decision- making, response and forecasting in crisis situations. Every few hours the State Fire Service, crisis management centres and the Institute of Meteorology and Water Management (IMGW) receive detailed data and reports.

This system makes a difference in crisis management as it offers precise, fast and reliable information. Since the beginning of the flood (Sept. 2024), the satellite monitoring system operated 24 hours a day. Until Friday morning (2024, Sept. 20) Civil Security Hub generated: 27 analyses of immediate effects of floods, 6 analyses of maximum water level, 112 dedicated views maps delivered within approximately 50 - minute processing time since the satellite's flight. CIK CBK PAN specialists also

[1] NATO defines Situational Awareness situational awareness /battlespace awareness as the knowledge of the elements in the battlespace necessary to make well-informed decisions. AAP-06(2013), 2-S-8.

APA Dictionary of Psychology defines situational awareness as conscious knowledge of the immediate environment and the events that are occurring in it. Situation awareness involves perception of the elements in the environment, comprehension of what they mean and how they relate to one another, and projection of their future states. In ergonomics, for example, it refers to the operator's awareness of the current status and the anticipated future status of a system. Situation awareness is influenced by a number of factors, including stress; it may be impaired by cognitive tunnelling or social tunnelling. https://dictionary.apa.org/situation-awareness.

use satellite images from the Copernicus program (Sentinel-1, Sentinel-2) and Landsat 8/9.

The recipients of the reports include: the State Fire Brigade, regional crisis management centers, hydrological service and the Operational Command of the Armed Forces. CBK PAN in cooperation with the Central Statistical Office and support for water pumping from IMGW also provides extended analyses with the assessment of damages.

This newly implemented satellite monitoring system illustrates the growing role of new technologies in crisis management. Poland's Civil Security Hub, equipped to respond to natural disasters, which will enable a more effective and coordinated response of entities. The centre will streamline the entire process by defining needs by identifying critical areas of observation, maintaining bilateral communication with multiple users and ensuring that information coming from space is delivered efficiently to meet their needs (Grochowalski, 2024).

4.3.3 Support

Support is organized in many forms by international, governmental as well as non-governmental organizations.

Financial help was offered by the European Union. Ursula von der Leyen while visiting a flood-damaged Wrocław, Poland met with the heads of affected countries—Poland, Austria, the Czech Republic and Slovakia. She promised that means will swiftly be transferred for repair from the EU's solidarity fund and pledged additional 10 billion euros for urgent repairs (EuroNews, 2024c, Sep. 19).

Ministry of National Education, Institute of Education Research (IBE), Centre of Education Development (ORE) and Warsaw Centre of Social-Educational Innovations and Training (WCIES) provide jointly support to schools, teachers, children and their families. The headmasters and teachers face challenges while managing schools in crisis situation and providing psychological aid to children and the youth. Teaching activities have been suspended in: Opole province–189 schools, Lower Silesia–220 schools, Lubuskie province-29 schools and Silesia province–1school. Ministry of National Education activated "Schools for Schools" platform which enables school communities to provide help to the flood-affected schools (Sokołowska, 2024).

Caritas Poland assigned 400 000 PLN for the emergency interventions and urgent help for the flood victims. Caritas organized also online donations.

Non-governmental organization Great Orchestra of Christmas Charity (WOŚP) donated 40 m PLN for the flood victims and organized the donation on its YouTube channel "Flood-Help".

Polish government has offered comprehensive financial solutions to people and entrepreneurs to ease the losses they suffered. They include:

- the solutions to ease the burden on borrowers, who lost their mortgaged homes as a result of the flood as well as solutions that will help to repay other loans,
- a regulation that will extend payment deadlines for VAT, PIT and CIT taxes for entrepreneurs affected by the floods,
- entrepreneurs affected by the floods can defer their Social Security contributions, as long as they declare that they have been affected by the flood,
- companies donating funds, products and services to flood victims can take advantage of a 0% VAT rate.
- an additional 1 billion PLN for aid and restoring infrastructure, bringing the grand total to 2 billion PLN from the state budget (Chancellery of the Prime Minister, 2024).

The Tauron Foundation donated PLN 2 million to purchase of dehumidifiers, pumps and generators and organized cash collections and transport of the most necessary items for the victims. Moreover, 1,000 employees were sent to the affected area, who secured electrical appliances and help the injured:

> More than 500 stations were flooded, low and medium voltage networks were damaged. It turned out that thousands of customers were without electricity. Our people went from house to house and checked if it was safe enough to restore delivery of electricity - Grzegorz Lot President of Tauron Polska Energia. (Huta, 2024, Oct.7)

The flood damaged 12 hydroelectric power plants in Lower Silesia (in Brzeg, Nysa and Bystrzyca Kłodzka). It will take weeks or months to restore them.

Moreover, Tauron Sales have launched a special offer for people affected by floods to make it easier to settle electricity bills during this difficult period. The customers can postpone the payment. The company do not to charge interest on customers struggling with floods, implement a fast service path on the hotline, launch the mechanisms for dividing arrears into interest-free instalments and postponing payment of receivables. In addition, there is a special proposal for flood victims, a price list with reference energy prices for Tauron customers, which will be in force until March 31, 2025. To simplify the procedures the only condition to benefit from the offer is to provide a decision number under which the municipality or social assistance centre granted the assistance.

4.3.4 Recovery

Rebuilding the flooded areas requires well planned, organized and cross-sectoral engagement of government, society and business. Government is preparing the recovery plan while The Polish Armed Forces are implementing the comprehensive operation "Phenix" (Wojsko Polskie. Operacja Fenix) from 23rd September as an element of the mission of supporting internal security and civilians within the civil-military cooperation framework (CIMIC) also with a strong focus on improvement and training. "Feniks" Operation as fully integrated scheme is organized and carried out along five lines of efforts:

1. Security - this line involves monitoring, protecting and reinforcing flood protection infrastructure and rebuilding damaged embankments and supporting police in protecting property on the flooded areas.
2. Health – this line involves providing medical aid, psychological support for the flood victims with PTSD symptoms, organizing medical evacuations (AIR MEDEVAC, CASEVAC) and facilitating vaccination to reduce the risk of infections. Taking over these tasks will enable the local medical teams to rebuilt their infrastructure.
3. Mobility – this means rebuilding bridges and restoring road patency and is a priority as it determines the quality of local life.
4. Logistics – this line concentrates on distributing energy and water to local communities to ensure the basic standard of living and facilitate the rebuilding processes.

5. Training – this line is about supporting local governments with necessary competencies and training programs to develop disaster risk management skills and strengthening local resistance.

4.3.5 *Impact*

A total of 8000 people (directly and indirectly) has been affected by the disaster at this point (IFRC) and over 20,000 buildings in the South of Poland have been impacted (ICEYE). These are numbers; however, flood means first of all human tragedy difficult to translate into measurable values. Damage of infrastructure, buildings and property is always the consequence of natural disaster impossible to avoid. The prediction for the nearest future is not very pessimistic: slight disturbances on the market, yet with no significant changes in GNP. The inflation may increase for a short time, the production and business activity especially in tourism, agriculture, transport and trade will slow down (Rutke, 2024). In the long run the impact of flood will be negative as the investments will be concentrated on rebuilding the damaged areas. Thus, investments and activities in other sectors will be delayed which will eventually negatively influence GNP.

4.3.6 *Lesson learned*

While discussing the future public policies the hydrologists indicate reduction of urban sprawl and resettlement from the flood prone areas as priorities to increase people's safety. Another aspect refers to infrastructure and building more dry reservoirs especially in the mountainous areas where they could intercept the excess water (Nowakowska, 2024).

Marian Banaś, the president of the Supreme Audit Office (NIK) reported that the main cause for flooding was lack of the special development plan for the affected areas. Local governments did not respect restrictions defined by Polish Waters (Wody Polskie) while issuing consents for land development to the investors. He also emphasized that delays in flood control investments and projects in Kotlina Kłodzka and The Nysa Kłodzka, Biała and Bystrzyca Dusznicka Basins directly contributed to the disaster (Wojak, 2024).

"Flood in the southern-western Poland shows as in a lens that establishing the effective civil protection system is a top priority. During last ten years it has been the next reactive levy in mass, which—although mass and partially effective—is difficult to coordinate and manage"—reads the InfoSecurity24 site. This opinion was formulated on the basis of the events that had occurred during the response operations. In Stronie Śląskie and Lądek Zdrój after the dramatic residents' appeal in social media and public Prime Minister Donald Tusk decided the chief of the Warmian-Masurian Fire Brigade to take over the responsibilities for disaster risk management from mayors of these cities who could not control the situation. However, there are presumably more villages, towns and cities where the local administration lacks competencies to manage crisis situations. Prime Minister appointed Marcin Kierwiński a plenipotentiary for implementing the flood recovery program. Moreover, on 9th November 2024 the Parliament approved the new Act on civil protection and civil defence which specifies the systemic foundations, procedures and training programs to prepare public administration and local governments as well as security services for future crises and shock, yet its implementation requires some time (Mikołajczyk, 2024). These actions demonstrated the culture of improvement and value for training.

The newly act on civil protection and civil defence is based on seven pillars:

1. Alarm systems. The civil protection bodies will be obliged to organize systems for monitoring threats, warning and alarming the population in the event of threats. The warning and alarm systems, like Galileo Emergency Warning Satellite Service[2] is to be created.
2. Evacuation and reception of the population. Preparing the plans for mass evacuations of the population and designating and preparing places of ad hoc dislocation.
3. Shelters. Securing financially projects of maintaining and building shelters and temporary shelters. Defining clear and safe criteria how new facilities are to be built.

[2] EWSS is an upcoming service which will be disseminating alert messages directly to the population of areas threatened by a looming natural or manmade disaster. The Galileo satellites will be transmitting to smartphones, or to any other navigation devices able to receive Galileo signals information related to the hazard (type of hazard, its severity, time of onset and expected duration, the targeted area) and instructions to follow. The alert content will be generated by national authorities and transmitted to Galileo for broadcast (https://defence-industry-space.ec.europa.eu/galileo-emergency-warning-satellite-service-underway-2024-01-24_en).

4. Strengthening social resilience. Social resilience will be built by increasing risk awareness, organizing training and exercises and instructing citizens how to behave in crisis situations. Social campaigns will serve as educational/information channels.
5. Building structures for civil protection and civil defence at various administrative levels (commune, county, province, country) and regulating maintenance of resources and training
6. Functioning of the civil protection system in times of war in the civil defence formula.
7. Financing of civil protection and civil liability insurance—minimum 0.3% of GDP covered from the budget of the Ministry of Interior and Administration and from the budget of the voivodes.

The civil protection will not be in the domain of the State Fire Service commander's responsibilities—as it was until now—but of the Interior Affairs Minister's. In the event of a state of war being declared, he will be the head of civil defence. The newly established Governmental Civil Protection Team will be an advisory and consultative body. This act sets also the hierarchy of bodies responsible for indirect protection of the populations, starting from those closest to the citizens: commune head (mayor, city president), mayor;voivodeship marshal, voivode, minister responsible for internal affairs, other ministers. The entities responsible for civil protection include: Government Security Centre, State Fire Service, volunteer fire brigades, services, inspections and guards, Government Agency for Strategic Reserves, enterprises providing basic services (e.g. city cleaning, water supply), social rescue organizations, healthcare entities, other entities. The National Rescue and Firefighting System will include a humanitarian aid component. Moreover, non-governmental organizations (e.g. the Polish Red Cross, Caritas and others), their resources and capabilities and their involvement in educational activities will be recognized and exploited. Under this act civil protection corps will be established. This is a team of selected and registered people who will be responsible for protecting the population in times of peace and for specific actions in times of war. The corps is to include representatives of civil defines personnel, national civil defines reserve.The Ministry of Interior and Administration and voivodes will be responsible for updating the created Classified Civil Défense Register (Horbaczewski, 2024). This newly created act on civil protection and civil defence addresses the present needs of the society determined by specific security conditions and nature of threats and challenges. It adapts the recommendations and good practices from the United Nation and The European Union and make the Polish system compatible with the international ones.

4.4 Evolution of Integrated Management – Discussion and Conclusions

The discussion on integration process of DRM system in Poland requires some reference to the Great Flood of 1997 when Poland experienced one of its most severe floods, caused by intense rainfalls. The rivers across the country overflew, inundating cities, towns and agricultural areas: the town of Kalisz on the Prosna River, Opole on the Oder River, from Wroclaw to Glogow along the Oder River, Nowa Sol, Kozuchow, Zielona Gora, Krosno and nearby Gubin (Guben) (European Space Agency). It was estimated that 54 people lost their lives, 1,100 villages were flooded in the southwestern part of the country, more than 500,000 acres of land submerged under flood waters, 140,000 people evacuated from their ruined homes.

The Anti-Crisis Committee for Fighting and Mitigation of Flood (ACC) established at national, provincial and local levels was responsible for co-ordinating evacuation, assistance to flood victims and maintaining order in inundated areas. Due to the inefficient information system the committees were able only partially process the hydrological information and forecasts, dispatch the necessary means and forces to the affected areas. Criticized by public opinion and media for their indolence and poor organization, the committees faced the challenge and coordinated rescue operations during and after the flood. The operations involved 25,000 firemen, 45,000 soldiers and 10,000 police officers, more than 100,000 civilians and volunteers altogether.

Media played two-faced. On one hand, TV programs (regular news, live coverages from the Anti-Crisis Committee meetings) were numerous and gained high viewer ratings. On the other hand, the destruction, panic and chaos in the affected areas of Poland at the beginning of the flood were presented against the "Ordnung" (German: order) of the preparatory action on the German side of the border. Actually, when the high water arrived at the Polish-German border reach and the Slubice-Frankfurt/Oder area, it appeared that the levees on the Polish side withstood the stress. On the German side, levees broke in several places causing massive inundations and high economic losses. The "alternative" weather forecasts, which grossly underestimated precipitation were published along the correctly foreseen in the official IMGW forecast. A local TV station in Warsaw disseminated information that a flood wave of 7 m height was approaching the capital city unprepared to respond to the crisis. Disinformation and mind games did not support the disaster management.

To respond to the needs of the affected population the government adopted the National Reconstruction and Modernization Programme (NRMP) for immediate, medium-term and long-term actions. The budget reserve for flood-related expenditures increased from USD24 million to USD 145 million. A package of assistance and assistance with fees, taxes, payment of interest on loans, preferential credits and indemnity for death was offered to flood victims. It should be noted that a majority of the people who suffered losses were not insured against the consequences of the flood (Zaleskiewicz et al., 2002). Referring to the classification by Coppola (2015) Poland fells into the mechanisms typical of the underdeveloped countries. This flood relief program has been the largest humanitarian action in the history of Poland supported by other countries and the European Union. Moreover, the support to people given by government and business of public sector (Tauron) described above after the flood 2024 shows that Poland still needs to introduce systemic reorganizations to the insurance mechanism.

The year 1999 was crucial for integrating disaster risk management system as the response to the flood of 1997 and governance changes. The new 3-level state administrative structure was established for better communication and disaster management and the maps of the flood threatened areas were drawn and made available to public. The flood of 1997 strengthened social awareness of risks, although there are some weaker points such as decreasing investments in flood control embankments in Poland from 1998 to 2023. Since 2015, the volume of embankments has significantly decreased and in 2023, the expenditures decreased by 64 per cent (Sas, 2024). In 2011 Poland gained 3.25 points in the Disaster risk reduction progress score (1–5 scale; 5 = best) close to World Median (World Bank Group).

The flood of 2024 was a test for the DRM system on how it has developed over the time and how it responses to the crisis situation within one framework integrating overall capacity (means, forces and skills).

Cross-functional collaboration was demonstrated at both phases response and recovery when for example, military forces and other formations local government, volunteers help to evacuate people or to build a field hospital in Nysa.

Communication, although distorted by fake news and the adversary's soft tactics was fluent, reliable and with regularly updated content thanks to the all engaged entities.

Technology integration improved the disaster management especially in the area of decision making; application the latest system of radar satellites to collect critical data for comprehensive risk assessment.

The exemplary strategic partnerships of Tauron Group focus on crisis management and the contributions of the energy groups to climate change mitigation: energy transition, biodiversity protection, investments in distribution infrastructure and energy storage to minimize the impact of failures and crisis power.

Culture of improvement and training programs are defined in the newly designed Act on civil protection as well as Phenix operation realized by the Polish Army.

It is worth mentioning also that under the agreement MI/DGWiŻŚ/2024/03/01/WF of 15 March 2024 (amended 15 Oct. 2024), Minister of Infrastructure granted to Polish Waters 507 496 997 PLN for activities within the framework of the Oder and Vistula Basins Flood Protection Project (Państwowe Gospodarstwo Wodne Wody Polskie).

The culture of improvement and integration strategy is evolving not only on the structural, organizational and management levels but also sectoral level joining two strategies of integrated flood management and natural flood management.[3] The pro-ecological anti-flood solution implemented by WWF Poland Foundation project *"Ecological flood protection in the Odra"*, improved the safety of the municipality of Wołów and

[3] *Integrated Flood Management* – the concept builds on the principles of the ecosystem approach by considering the entire basin ecosystem as a unit; promotes an integrated land and water resources development in a river basin, and aims at maximizing the net benefits from the use of floodplains and minimizing loss of life from flooding. This involves participatory, cross-sectoral and transparent decision-making and defining the best mix of strategies. The Associated Programme on Flood Management. (2004). Integrated Flood Management.

Natural Flood Management works through managing soil, wetlands, woodlands and floodplains to retain water at times of flood risk. It brings benefits to biodiversity, climate change mitigation, and water quality, increasing the resilience of communities and landscapes to the challenges of climate change, at the same time reducing the conflict between

supported the restoration of valuable forest habitats. As Piotr Nieznański form WWF Poland Foundation says:

> In 2020, there was a flood for the first time and this solution was successful. Restoration of floodplains resulted in a decrease in the flood wave and increased the retention. New shafts, located further from the river, are safer and better protect residents. (WWF, 2021)

Dams and embankments very often do not withstand, whereas the river can deal with the excessive water itself. Therefore, the most effective, the cheapest and durable method of flood protection is to give the river on certain sections its natural space. Widely spread floodlands, wetlands, riverside meadows and forests act as a sponge, slowing down and lowering the flood wave. When the river floods it does not damage the riverside areas and supports their natural functioning (WWF, 2021).

The public debate on flood protection is now even more wide and engaged scientists, practitioners and experts. They emphasise the need for more comprehensive approach ranging from effective alert systems for the people to nature-based solutions, as setting infrastructure against nature-based solutions is misleading and dangerous. The reservoirs reduce a flood wave, yet the delay between rainfall and runoff depends also on the topography of the catchment area, its geological structure and soil type, as well as land use and the condition of rivers in the area.

It is possible to avoid small disasters but not a disaster of that strength and scale. The experience of the flood 2024 and the previous ones shows that the plans how to predict and mitigate the natural disasters affecting the region should be drawn from the comprehensive cross-sectoral perspective, combined engineering and natural solutions and executed in the coordinated (integrated) systemic manner (Kość, 2024). Extreme weather events require combined actions on climate change adaptation and disaster risk reduction and both need to embrace the whole risk management cycle: prevention, preparedness and recovery (Climate ADAPT).

private property and river dynamics and the resulting risks to life and damages. Moreover, managing river basins implies an international coordination and cooperation across functional and administrative boundaries. Murray (2017).

References

Alcántara-Ayala, I., Thouret, J.-C, Geertsema, M., Cienfuegos, R., & Cui, P. (2022). Editorial: Integrated disaster risk management: From earth sciences to policy making. *Frontiers in Earth Science, 10*, 1017489. https://doi.org/10.3389/feart.2022.1017489. https://www.frontiersin.org/journals/earth-science/articles/10.3389/feart.2022.1017489/full

Arnold, M. (2012). In B. Wisner, J. C. Gaillard, & I. Kelman, *International planning systems for disaster*.

Bischoff, Sh. (2021). Integrate your organization say goodbye to silos. Be More Unified. Nonprofit Impact, www.nonprofitimpact.com.

Brzozowski, A., & Rybak, M. (2024, Sep. 19). *ABW złapała fałszywego wojskowego, który straszył wysadzeniem wałów pod Wrocławiem. Podszywał się pod kontrwywiad*. https://wroclaw.wyborcza.pl/wroclaw/7,35771,31318015,abw-poszukuje-falszywego-wojskowego-ktory-straszy-ludzi-wysadzaniem.html

Business Insider. (2024 a, Sep.18). *Stan klęski żywiołowej w Polsce. Zagrożone obszary [MAPY]*. https://businessinsider.com.pl/wiadomosci/powodz-2024-stan-kleski-zywiolowej-w-polsce-zagrozone-obszary-mapy/y538z2y

Business Insider. (2024 b, Sep.19). *W zbiorniku Racibórz Dolny coraz mniej wody. Najnowszy raport*. https://businessinsider.com.pl/wiadomosci/w-zbiorniku-raciborz-dolny-coraz-mniej-wody-najnowszy-raport/yk9ls7z

Celej, P. (2024, Sep. 18). *Już 14 tysięcy żołnierzy walczy z powodzią i jej skutkami na południu Polski*. https://www.gazetaprawna.pl/wiadomosci/kraj/artykuly/9609035,juz-14-tysiecy-zolnierzy-walczy-z-powodzia-i-jej-skutkami-na-poludniu.html

Chancellery of the Prime Minister. Republic of Poland. (2024, Sep. 18). *Non-refundable subsidies for homeowners with mortgages*. https://www.gov.pl/web/primeminister/non-refundable-subsidies-for-homeowners-with-mortgages

Coppola, D. P. (2015). The management of disasters. In *Introduction to international disaster management* (3rd ed.). Science Direct

ClimateADAPT. EU. Disaster risk reduction. (2024). https://climate-adapt.eea.europa.eu/en/eu-adaptation-policy/sector-policies/disaster-risk-reduction

EuroNews. (2024 a, Sep. 17). *Poland declares state of natural disaster in areas affected by severe flooding*. https://www.euronews.com/my-europe/2024/09/17/poland-declares-state-of-natural-disaster-in-areas-of-flooding

EuroNews. (2024 b, Sep. 19). *Rising rivers threaten southern Poland as flooding recedes elsewhere in Central Europe*. https://www.euronews.com/my-europe/2024/09/19/rising-rivers-threaten-southern-poland-as-flooding-recedes-elsewhere-in-central-europe

EuroNews. (2024 c, Sep. 19). *EU Commission chief visits flood-stricken region in Poland and pledges billions in swift aid*. https://www.euronews.com/2024/09/19/eu-chief-visits-flood-stricken-region-in-poland-and-pledges-billions-in-swift-aid

European Space Agency. (1997). *Flood in Europe (Poland), 1997.* https://earth.esa.int/eogateway/gallery/flood-in-europe-poland-1997

Giorgou, E., & Zeith, A. (2023, Apr.20). When the lights go out: the protection of energy infrastructure in armed conflict. *Humanitarian Law and Policy.* https://blogs.icrc.org/law-and-policy/2023/04/20/protection-energy-infrastructure-armed-conflict/

Goh, M. H. (2024, May 14). [CM] The concept of integrated crisis management. *BCM Institute.* https://blog.bcm-institute.org/crisis-management/the-concept-of-integrated-crisis-management

Greenberg, M. (2017, Feb.). The city at war. *International Journal of Urban and Regional Research.* https://www.ijurr.org/spotlight-on/the-city-at-war-reflections-on-beirut-brussels-and-beyond/the-city-at-war/

Grochowalski, P. Z. (2024, Oct. 14). *System monitoringu satelitarnego—powódź 2024.* https://www.urania.edu.pl/wiadomosci/system-monitoringu-satelitarnego-powodz-2024

Horbaczewski, R. (2024, Sep.17). *Projekt ustawy o ochronie ludności przyjęty przez Radę.* Prawo.pl.https://www.prawo.pl/samorzad/ustawa-o-ochronie-ludnosci-nowe-zalozenia/525482.htmlhttps://www.prawo.pl/samorzad/ustawa-o-ochronie-ludnosci-nowe-zalozenia/525482.html

Huta, I. (2024, Oct.7). Tauron group on key activities and strategies during the recent floods in southern Poland. PRECOP 2024. *Dziennik Zachodni.* https://dziennikzachodni.pl/grupa-tauron-o-kluczowych-dzialaniach-i-strategiach-podczas-ostatnich-powodzi-w-poludniowej-polsce-precop-2024/ar/c1-18852187

ICEYE. (2024, Sept.). Flooding from Storm Boris in Central and Eastern Europe. https://www.iceye.com/solutions/storm-boris-september-2024

IFRC. *(2024, Sept.). POL: Flood - 09–2024 - Flood in Poland 2024.* https://go.ifrc.org/field-reports/17251

Integrated flood Management. (2009). Concept paper. World Meteorological Organization, WMO-No-1047.

Kość, W. (2024, Oct. 24). In wake of devastating floods, Poland debates how to withstand the next deluge. *Notes from Poland.* https://notesfrompoland.com/2024/10/24/in-wake-of-devastating-floods-poland-debates-how-to-withstand-the-next-deluge/

MedExpress. (2024, Sep. 20). *Flood 2024: If someone needs help, that medical help is already there.* https://www.medexpress.pl/en/events-campaigns/flood-2024-if-someone-needs-help-this-medical-help-is-now/

Mikołajczyk, D. (2024, Sep. 19). *Powodziowe pospolite ruszenie, czyli kryzys w polskim wydaniu.* https://infosecurity24.pl/bezpieczenstwo-wewnetrzne/powodziowe-pospolite-ruszenie-czyli-kryzys-w-polskim-wydaniu

Murray, A. (2017), *Natural flood management adopting ecosystem approaches to managing flood risk.* Report for Friends of the Earth.

Nowakowska, N. (2024, Sep. 20). Hydrolog nie ma złudzeń. Tak powinno zmienić się mieszkalnictwo po powodzi. *Interia*. https://wydarzenia.interia.pl/kraj/news-hydrolog-nie-ma-zludzen-tak-powinno-zmienic-sie-mieszkalnict,nId,7819225#utm_source=paste&utm_medium=paste&utm_campaign=firefox 5

Pakuła, D. (2024, Sep. 19). Rząd uruchamia operację FENIKS. Każdy będzie zaopiekowany. *Interia*. https://wydarzenia.interia.pl/opolskie/news-rzad-uruchamia-operacje-feniks-kazdy-bedzie-zaopiekowany,nId,7818535#utm_source=paste&utm_medium=paste&utm_campaign=firefox

Państwowe Gospodarstwo Wodne Wody Polskie. Zadania finansowane lub dofinansowane z budżetu państwa lub z państwowych funduszy celowych. (28 March 2024 r). https://www.gov.pl/web/wody-polskie/zadania-finansowane-lub-dofinansowane-z-budzetu-panstwa-lub-z-panstwowych-funduszy-celowych

Rutke, M. (2024, Sep. 17). Powódź i stan klęski żywiołowej w Polsce. Oto skutki dla gospodarki. *Business Insider*. https://businessinsider.com.pl/gospodarka/powodz-i-stan-kleski-zywiolowej-w-polsce-oto-skutki-dla-gospodarki/2b8jhlf

Sandoval, V., Voss, M., Flörchinger, V., et al. (2023). Integrated Disaster Risk Management (IDRM): Elements to advance its study and assessment. *International Journal of Disaster Risk Science, 14*, 343–356. https://doi.org/10.1007/s13753-023-00490-1

Sas, A. (2024, Sep. 27). Built and modernized flood control embankments in Poland from 1998 to 2023 (in kilometers). *Statista*. https://www.statista.com/statistics/1493779/poland-built-and-modernized-flood-barriers/

Science for Disaster Risk Management. (2020). In A. C. Valles, M. M. Ferrer, K. Poljansek, & I. Clark (eds.), EUR 30183 EN, Publications Office of the European Union, Luxembourg, 2021, ISBN 978–92–76–18182–8, https://doi.org/10.2760/571085, JRC114026.

Sendai Framework for Disaster Risk Reduction 2015-2030. UNDRR.

Sokołowska, M. (2024, Sep. 19). *Ponad 400 placówek zamkniętych. Jest komunikat ministerstwa*. https://wydarzenia.interia.pl/kraj/news-ponad-400-placowek-zamknietych-jest-komunikat-ministerstwa,nId,7818780#utm_source=paste&utm_medium=paste&utm_campaign=firefox

Starcevic, S. (2024, Sep. 19). *Tusk jokes that poles shouldn't 'panic' if they see German soldiers helping out after floods*. https://www.politico.eu/article/donald-tusk-jokes-poles-shouldnt-panic-german-soldiers-helping-floods/

Stezycki, K., Laizans, J., & Stoklasa, R. (2024, Sep. 18). *Central European floods leave trail of devastation; new areas to evacuate*. https://www.reuters.com/world/europe/poland-fortifies-towns-deadly-floods-afflict-central-europe-2024-09-17/

The Routledge Handbook of Hazards and Disaster Risk Reduction. (2012). London/New-York, pp. 603 - 619; https://www.preventionweb.net/hubs/walk-though-history-disaster-risk-reduction

TVP World. (2024, Sep. 15). *Heavy rain continues causing widespread flooding in Poland and Central Europe*. https://tvpworld.com/82313397/heavy-rain-continues-causing-widespread-flooding-in-poland-and-central-europe

UNDRR. (2019). A walk through DRR history. https://www.preventionweb.net/hubs/walk-though-history-disaster-risk-reduction

WHO. *Communication*. (2025). https://www.who.int/emergencies/risk-communications

Wojak, A. (2024, Sep. 23). *Ostatniej powodzi dało się uniknąć? "Brak inwestycji spowodował, że mamy do czynienia z tragedią*. Fakt. https://www.fakt.pl/polityka/zaniedbania-w-inwestycjach-przeciwpowodziowych-nik-bije-na-alarm/ssk4bhq

Wojsko P., & Operacja, F.(2024). https://www.wojsko-polskie.pl/articles/tym-zyjemy-v/2024-09-204-operacja-feniks/

World Bank Group. (2011). Disaster risk reduction progress score (1–5 scale; 5=best). *Prosperity Data 360*. https://prosperitydata360.worldbank.org/en/indicator/WB+CC+EN+CLC+DRSK+XQ

WWF. (2021, Feb. 19). *Ochrona przeciwpowodziowa i ochrona przyrody. to połączenie jest możliwe!* https://www.wwf.pl/aktualnosci/ochrona-przeciwpowodziowa-i-ochrona-przyrody

Zaleskiewicz, T., Piskorz, Z., & Borkowska, A. (2002). Fear or money? Decisions on insuring oneself against flood. *Risk, Decision and Policy*, *7*(3), 221–233. https://doi.org/10.1017/S1357530902000662

CHAPTER 5

Education and Training—Design Thinking for DRM

Abstract People are rather well educated and organized to face simple disasters, yet, they are much less competent when facing complex risks in the interconnected systems that build the structures of our world, such as organizations, economies, societies and the environment. Thus, effective preparedness requires building social capacity through education and training. The chapter present the design thinking as the teaching approached tailored for the security needs. The theoretical theories supporting this choice as well as methodologies and practical implementation in short/agile courses are presented and skills frameworks analysed as the guidelines for security education and training programs.

Keywords Design thinking · Agile project management · Capacity building · Teaching/learning theories

5.1 Educating for Security

Education, research and innovation have become the widely promoted tools by policies and practices for disaster risk reduction at national and local levels. As The Sendai Framework for Disaster Risk Reduction 2015—2030 reads:

S. Kowalkowski et al., *Cities in Crisis*,
https://doi.org/10.1007/978-3-032-08843-7_5

> [...] in order to reduce disaster risk, there is a need to address existing challenges and prepare for future ones by [...] cultural and educational resilience of persons, communities and countries and the environment, as well as through technology and research (p. 11).

Similarly, *The ROADMAP 2021-2030 for a disaster-resilient European and Central Asian region by 2030* that facilitates *The Sendai Framework for Disaster Risk Reduction 2015–2030*, coveys the notion of resilience as a "public good" which may be created by application of multi-stakeholder, all-of-government and all-of-society approaches that value diverse perspectives and contexts of resilience; inclusive understanding and multi-stakeholder investments; assign the roles and responsibilities for risk reduction among citizens, government, municipalities and the private sector; engage multiple sectors incorporating expertise on biodiversity, climate modelling, disability, gender equity or youth-related issues; engage science and technology stakeholders in inform evidence-based planning and risk reduction for investments; whereas knowledge management systems facilitate exchanges and shared learning between different ministries and relevant stakeholders.

People are rather well educated and organized to face simple disasters, yet, they are.

much less competent when facing complex risks in the interconnected systems that.

build the structures of our world, such as organizations, economies, societies and the.

environment (World Economic Forum Geneva, 2018, p.9). Thus, the top priority for all societies and their governments should be developing awareness of all aspects of natural disasters and find solutions that better mitigate the risk and fallout of the devastating events (Bevere, 2018).

Consequently, disaster risk management knowledge should embrace the full disaster management cycle (prevention, mitigation, preparedness, response, recovery and rehabilitation) and be based on an understanding of disaster risk in all its dimensions of vulnerability, capacity, exposure of persons and assets, hazard characteristics and the environment (CRED, UNISDR, 2016). Capacity building is "the process by which individuals, organizations, institutions and societies develop abilities to perform functions, solve problems and set and achieve objectives. It needs to be addressed at three inter-related levels: individual, institutional and societal" (United Nation Economic and Social Council, 2006). Thus,

Disaster Risk Management (DRM) capacity building can be defined as "efforts to strengthen the competencies and skills of a target organization, group or community so that the target could drive DRR efforts, or in a broader sense development, in a sustainable way in the future" (International Federation of Red Cross and Red Crescent Societies, 2013).

Comprehensive School Safety Framework 2022-2030 for Child Rights and Resilience in the Education Sector (GADRRRES, 2022) may serve as the embodiment of the UN and EU concepts and policies for disaster risk reduction. School sector (community) faces multiple hazards, including natural disasters, technological threats, climate change, epidemics, violence, conflict and everyday threats which will impact the development of children and societies. The comprehensive school safety proposes a participatory risk-informed approach to realize three main goals:

1. Protecting students and staff from death, injury, violence and harm.
2. Planning for education and protection continuity and limit disruptions to learning when shocks or threats of all kinds occur.
3. Promoting knowledge and skills of students and duty-bearers, to contribute to risk reduction, resilience building and sustainable development.

On the academic level the example of the security education program is *The Academic Legion* in Poland. It is a voluntary military training of students and graduates carried out in universities, military units and training centres of the Polish Armed Forces. The Academic Legion program is a part of basic training (voluntary basic military service) and the non-commissioned officer and officer modules (Zostań Żołnierzem).

The disaster risk knowledge should be incorporated in formal and non-formal education, as well as in civic education at all levels, as well as in professional education and training; The teaching content should relate to life skills, risk reduction, safety, climate change action, health and hygiene and social cohesion.

The methods applicable for promoting public education on disaster risk reduction should foster peer-to-peer awareness-building, leadership, cultural expression and.

psychosocial support, taking into account specific audiences and their needs. The widely exploited methods include campaigns, social media and community mobilization and regular courses mostly online.

The literature and good practice provide several approaches for education: Design Thinking (DT), Project Based Learning (PBL) and Understanding by Design (UbD). The differences between them are revealed in a teaching process and evaluation as well. Project-based learning (PBL) is a category of educational activities in which curricular activities revolve around students' engagement in a problem or project. PBL divides a study into a series of discrete activities using student-directed learning. A PBL aligns with curricular assessment timeline with testing, examination, inquiry and production of outcomes.

The Understanding by Design (UbD) approach also uses creative problem solving as the central learning activity, yet differs in evaluating results. In UbD, teachers help students define the problem and develop a final result. In some courses with complex and demanding topics, there may not be a readily available answer (Jia et al., 2023; celt.iastate.edu.).

Design Thinking is intentionally more open-ended and then better suited for complex topics (Hokanson, Nyboer, 2018). As a part of the broader project-based learning concept, it uses a creative, systematic approach to teach problem-solving to better meet the expectations of present realities.

5.2 Teaching/Learning Theories Supporting Design Thinking and Agile Methodologies in Education

The theories which support the concept of design thinking in education are behaviourism, cognitivism and connectivism.

Behaviourists Watson and Skinner claim that if they were given a group of infants, the way they were raised and the environment they put them in would determine how they acted, not their parents or their genetics. The things we experience and our environment are the drivers of how we act. Thus, knowledge is a repertoire of behaviours: "knowledge is action, or at least rules for action" (Skinner, 1976, 152), e.g.: "I know a bluebird when I see one" can be an equivalent to "I have the capacity to identify a bluebird although I am not now doing so" (154). Knowledge is construed as a repertoire of behaviours and someone understands something if they possess the appropriate repertoire.

5.2.1 Behaviourism Learning Theory

The stimulus-response sequence is a key element of understanding behaviourism. Even complex actions can be broken down into the stimulus-response.

Students are a passive participant in behavioural learning—teachers are giving them the information as an element of stimulus-response. Teachers show students how they should react and respond to particular stimuli. This activity should be repeated to regularly remind students what behaviour is expected. The exercises behavioural learning relies on include: “skill and drill” which provides the consistent repetition necessary for effective reinforcement of response patterns, question (stimulus) and answer (response) frameworks in which questions are of gradually increasing difficulty; guided practice; and regular reviews of material.

Motivation plays an important role in behavioural learning. Positive and negative reinforcement can be motivators for students. For example, a good mark on a test can serve as a positive reinforcement. A bad mark is a negative reinforcement, yet it is also motivation to get the student to act in a desired way (WGU, 2020).

Behaviourist teaching methods have proven most successful in areas where there is a “correct” response or easily memorized structured material: facts and formulae, scientific concepts and foreign language vocabulary, whereas its application in teaching comprehension, composition and analytical abilities is disputable (WGU, 2020).

Behaviourism gave foundations for Behaviourism Design (BCD) an approach to build breakthrough solutions to challenges in any area of human activity. It is successfully applied for creating new products, solutions which influence desired human behaviours.

Practical implementation of the concept can be illustrated by the methodology of RARE—Centre for Behaviour and Environment.

The Behaviourism Design steps respect the ideas of diversity, equity and inclusion (DEI). Diversity—recognizing all the ways that people differ and are unique:

Equity—creating fair treatment, opportunities for all people and identify and eliminate barriers that have prevented the full participation of a particular group.

Inclusion—creating environments in which any individual or group is respected, supported and valued.

The traditional, commonly applied levers to change behaviour are: information, rules and regulations and material incentives. They are generally effective, yet they can also produce the undesirable side-effects. Information does not always lead to action, incentives can backfire or send the wrong message and rules can be difficult to enforce. Behavioural science provides deeper insights into human's motivations and actions: people make decisions based on their emotions, what other people are doing and expecting them to do and how we structure the context for decision-making.

The toolkit of Behaviour-Centred Design includes emotional appeals, social influences and choice of architecture to refer to people's full range of behavioural motivations and barriers.

1. Information

Providing information about what the target behaviour is, why it matters and how to do it.

by step-by-step instructions, training on the target behaviour.

Build awareness and understanding at forums, meetings, provide feedback on performing the target behaviour

2. Rules

Enacting rules that promote or restrict a behaviour

3. Material incentive

Make the target behaviour more convenient and accessible to do (e.g. remove barriers, provide substitutes).

Make the undesired behaviour more difficult to do (e.g. create friction points, barriers).

Give rewards or penalties

4. Emotions

Leverage emotions in specific contexts

- Pride: to motivate people to boast about their reputable action/ behaviour
- Hope: to motivate people to make an attempt to achieve a desired outcome while facing a threat/problem
- Fear: to motivate people to avoid risks when in a threatening situation
- Anger: to motivate people to confront others when witness or are involved in a situation against their values
- Interest: to motivate people to search for information when necessary
- Prospect of shame: to motivate people to avoid an action when others might find out about socially-undesirable actions.

Personalize the message

- Be appealing, focus on a single story over abstract statistics
- Make a campaign personally relevant, relatable and appealing

5. Social influence

Make engaging or not engaging in the target behaviour observable.
Make the target behaviour the perceived norm.
Eliminate excuses for not engaging in the target behaviour

6. Choice of Architecture

Changing the context in which choices are made.
Direct attention.
Simplify messages and decisions.
Use timely moments and prompts.
Facilitate planning and goal setting.
Use commitments to bind or limit future decisions (Rare).

The methodology consists of 8 steps and can be applied in the area of security for solving social, political, cultural, environmental problems as illustrated in Fig. 5.1.

Cognitivism developed as a critique of a behaviourist approach to learning. It concentrates on information processes as the foundation of learning—how a person's mind receives, organizes, saves and retrieves information. According to cognitivism, the learner employs different types of memory during learning and knowledge acquisition. A learner receives information transmitted through the senses to the brain, this information remains for a second in the sensory store and if it is not transferred to active memory immediately, it is lost. Therefore, educators are expected to create for learners Advantageous Learning Environments that facilitate connections between new information and information already in long-term memory, e.g.: when learners listen to thought-provoking questions, it tutors their brains to move further from their current knowledge to get solutions. Cognitivists believe that humans learn from thinking and experience rather than from presented facts and procedures. Learning is an internal process not a product and is acquired when learners actively engage in problem-solving activities (Schunk, 2012, Structural Learning com. 2022). Online education recognizes the importance of individual

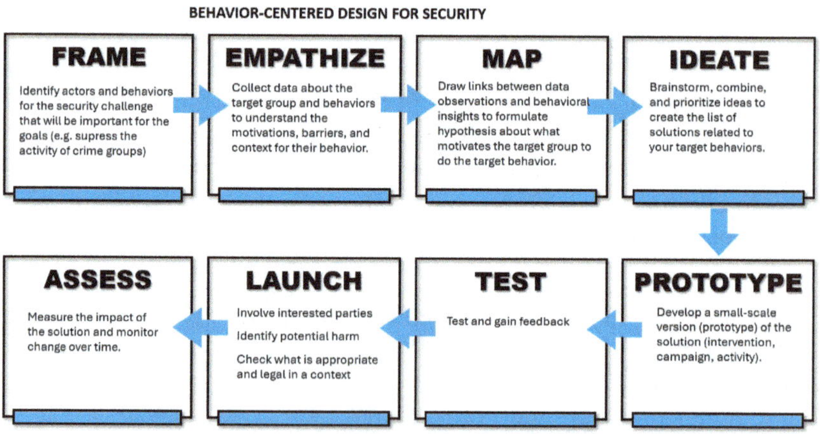

Fig. 5.1 Behaviour-centred Design for Security. *Source* based on: Rare. Centre for Behaviour &The Environment. *Behaviour Design* https://behavior.rare.org/behavior-centered-design-landing/

difference in learning as it supports learners' selection of the best methods for recalling existing learning before moving to new cognitive structure. Online education offers different learning styles— variations in perception, interaction and response to the learning environments. Students who learn more effectively through concrete experience are given opportunities to work with peers and get peer feedback, with the instructor who is less an authority and more a coach. Learners who prefer reflective observation and see the instructor as an expert can select learning processes that emphasize reflection (e.g.: journaling, descriptive writing, conversations with a group of people and pausing to absorb and digest information). Those who learn best through abstract conceptualization can be given opportunities to research a model or theory. Learners who prefer active experimentation can take part in activities that apply information to real life (SAGE Encyclopaedia).

Connectivism is built on the assumption that the process of Meaningful learning occurs in a network environment that is interactive and varies in intensity and levels (Davidson & Glassner, 2015). Digital technology brings people together and creates new learning and training opportunities and transforms workplace training practices. It requires applying LMS tools that offer a social component. The intersection between human interaction and digital technology supports a continuous learning environment (Davidson, Glassner, 2015). George Siemens (2004) and Stephen Downes (2005) claim that connectivism begins when an individual uses a digital technology to solve a problem, e.g.: googling a question, texting a friend, or searching for topical social media content. This way they gain an insight into a particular problem or topic. George Siemens elaborated eight guiding principles for Connectivism Learning Theory:

1. **Learning and knowledge rest in a diversity of opinions.** Perspectives from a variety of sources deepen our insight.
2. **Learning is a process of connecting specialized nodes or information.** Building relationships with colleagues, creates the opportunity to gain or develop new skills, thoughts and ideas we might not otherwise have access to.
3. **Learning may reside in non-human appliances.** Information may be stored in a digital form, in an application, social media post, or video. A community of learners may store information in a database or forum.

4. **Capacity to know more is more critical than what is currently known.** In Siemens' words: "Our ability to learn what we need for tomorrow is more important than what we know today".
5. **Nurturing and maintaining connections are needed to facilitate continual learning.** Collaborative social interaction brings people together and creates a long-term learning environment.
6. **Ability to see connections between fields, ideas and concepts is a core skill.** It is crucial to learn how to bridge point A and point B. That bridge itself is a new learning opportunity.
7. **Accurate, up-to-date knowledge is the intent of all Connectivism Learning activities.** When working together, our understanding is being reinforced and updated.
8. **Decision-making is itself a learning process. Choosing what to learn and the meaning of incoming information is determined by a shifting reality** (Picciano, 2017, 360Learning).

5.3 Design Thinking—The Concept and Practice

Design thinking (DT) was popularized by the Silicon Valley design firm IDEO and its applicability to a wide range of challenges and solutions. Design thinking is integrated with science to solve complex, real-world problems and introduce significant societal changes. Design thinkers strive to balance what is desirable from a user's point of view with what is feasible with technology and viable from a business factors perspective (IDEO).

There are five principles behind the approach and the process that brings these principles to life (Morris and Warman, 2015, Matthews and Wrigley, 2017)[1]:

1. Empathize: The best designs are human cantered. Human beings (customers, soldiers, victims) and their needs at the centre of the

[1] Harvard Business School proposes the 4-phase design thinking procedure for organizations: Clarify: the team workers identify the problem that will be explored to ensure the best possible outcome; Ideate: they generate original, innovative ideas for solutions; Develop: they experiment with the solutions they conceived in the ideate phase. Prototypes should not be considered "final," but rather as objects to test and learn from; Implement: they test each prototype. The goal is to learn and collect data necessary to further iterate on potential solutions (Stobierski 2022).

process helps rather than a specific technology conditions or context leads to innovative solutions.

2. Define: Framing the problem is the foundation to the design. Asking the right question(s) involves the users early on to seek user evaluation of a concept.
3. Ideate: Innovation is born from the class of ideas. The best solutions are obtained by transforming ideas from different sources to fit the particular context.
4. Prototype: Showing is more effective than telling. Experiencing solutions (learning from a sample) is far more engaging and illuminating than describing them.
5. Test: Live testing and iterating bring innovations to life. (compare with Behaviour-cantered desig fig.)

5.3.1 *Empathize*

While seeking to understand users'/people/the following methods may prove useful: observational research, ethnographic methods, watching, listening, discussing and seeking to understand, surveys and focus groups. However, surveys and focus groups may provide only superficial knowledge and actually they may block the development of empathy. Sometimes, what people say they do and what they really do can be different, for example, when students asked about their study habits in a focus group, they often describe their intentions rather than their actual behaviours. The way to avoid this tendency is to ask for specific examples about extreme scenarios, for example, to understand students' experience with school, one may ask: "What was the absolute best experience you had in a class?" or "What was the worst experience you ever had studying for a test or working on a particular assignment?" These extreme examples can provide insight into how people truly respond to the conditions of their environment and lessens the influence of the social norms requirements.

Another challenge with these tools is formulating questions. The most relevant ones will be asked usually when people are observed in the natural or original position or place. For example, when observing students studying during their subway commute, the questions specific to the environment's challenges will occur: how they contend with the noise and crowding and how they take notes while standing in a moving train.

Observing people in context helps to ask better questions and provides data about thoughts and feelings that really motivate them.

User Experience (UX) is critical to the success of a product or service. As Paul Boag states: "To be a great designer, you need to look a little deeper into how people think and act", we believe that this applies to everyone, who creates and often executes plans for any projects: security, rescue, humanitarian ones, etc. There are seven factors that influence user experience (Morville, 2004):

1. Usefulness—the product is useful if it serves some purpose, not necessarily practical one. It can deliver benefits such as fun or aesthetic appeal. For example: art gallery visitors may 'use' the sculpture to educate themselves and/or gain spiritual pleasure (art therapy courses).
2. Usability—this is a feature which enables users to achieve their end objective. Poor usability is often associated with the very first generation of a product, e.g. the first generation of MP3 players. The usability may be judged by five criteria: learnability—how easy is it for users to deal with the product the first time they use it?; efficiency—can the user complete their desired tasks with a product quickly/easily?; memorability—can users recognize functionality of the product after some time of not using it?; errors—how many errors do users commit when using the product and how severe are these errors?; satisfaction—does it provide a satisfactory experience, or a frustrating one? (Nielsen, 1993; Stevens, 2024). In other words, user-friendliness and ergonomics have become important quality characteristics for products. For example, *IEC 62366* is the international standard that covers the application of usability engineering to medical devices. The nanomaterials demonstrate enhanced usability as they can have greater chemical reactivity, increased control of the light spectrum, lighter weight and higher strength as compared to their larger-scale counterparts, e.g. waterproof smartphones (in Asia, shops with special machines will take a smartphone and make them go through a cleaning process of 20 minutes and the application of nanocoating) (Nanografi Explains). Nanotechnology in the military will improve medical and casualty care for soldiers and enable production of lightweight, strong and multi-functional materials for use in clothing, both for protection and to provide enhanced connectivity. It is projected that the medical nanobots and

nano-enhanced reconnaissance and communication devices (such as micro-radar for miniature vehicles) will be used from 2030 onwards (Soutter, 2023).

3. Findability—If the content is difficult to find the customer will certainly abandon searching for it. In the extreme, life-threatening situations easy/fast access to a product (e.g. medicine, water, etc.) is critical.
4. Credibility—a product is credible to users if it is transparent about how it uses users' data, includes testimonials and reviews and basic data about the producer.
5. Desirability—is a feature that considers users' feelings and emotions.
6. Accessibility—means inclusive design that not only meets the needs of people with disabilities but creates experiences that are usable by everyone, irrespectively of their age, gender, ethnicity, culture, or other characteristics.
7. Value—a product is valuable if it solves users' problems and delivers a return on investment. For example, food delivery apps became extremely valuable to people during lockdowns. They kept many restaurants operating while providing customers with meals (https://www.uxpin.com/studio/blog/ux-honeycomb-definition-and-use/).

5.3.2 *Framing the Problem*

The next step in design thinking is to frame the problem a critical step in the whole process as the point where and how you start a design challenge has much to do with where you end up it. A great design challenge starts with a user focus. It should also be broad enough to give us the opportunity to innovate. Three simple questions may help frame the problem:

- Who is your user and what benefit are you trying to provide? (user-centred approach).
- Why do you want to do this? (broadening the frame).
- What's preventing you from doing it? (narrowing the frame to a reasonable starting point).

Projects often start with misguided assumptions that fall away once these three questions are seriously considered. The next step is ideation, or *brainstorming*. In this phase of the design thinking process, quantity is far more important that quality. "Bad" ideas often give rise to good ideas when they juxtaposed with other ideas and new scenarios are shaped from modifying them.

Brainstorming sounds simple, but it's not easy. We usually present small, safe ideas and spend a lot of time debating on them and finally produce something that worked before. It is important to move away from safe and already known ideas to propose variety brave ones. There are some IDEO's rules for such brainstorming: defer judgment (on your own ideas and those of others), encourage wild ideas, build on the ideas of others, stay focused on the topic, have one conversation at a time, be visual (i.e. find a way to visually express your ideas) and go for quantity. The crucial nuance is calling the session a "brainstorm" to communicate that it is time to face a challenge and assemble a diverse group to collaborate on possible solutions.

5.3.3 Prototyping

Typically, in the ideation/brainstorming phase ideas are written on Post-it notes in a few words. In the prototyping phase, they are expanded, taken beyond the comfort zone and shown to a potential user through the step-by-step scenarios or a physical mock-up of a product or program to address a user's needs.

5.3.4 Conveying Ideas

The idea can be expanded beyond the Post-it note by storyboards as they help create a shared understanding of the story; expose story flaws and encourage to explore key aspects of the story. Other useful tools are sketches or handmade constructions, role playing and acting out user interactions, with video.

5.3.5 Testing

In a testing phase the prototype is exploited in a live environment. The invited real users can be observed for the feedback. It is worth emphasizing following Morris and Warman's words that: design thinking in its testing phase does not use the scientific method to verify if something is false or true. A design thinking experiment is "an attempt to grow your idea and figure out what you must do to ensure the idea's ultimate success" (Morris & Warman, 2015).

> All great innovators in literature, art, music, science, engineering and business have practiced it. So, why call it Design Thinking? Well, that's because design work processes help us systematically extract, teach, learn and apply human-cantered techniques to solve problems in a creative and innovative way—in our designs, businesses, countries and lives (Interaction Design Foundation).

Students like Silicone Valley designer progress through the stages of DT in search of innovative solutions to solving problems. In DT teachers *do not* define the problem, only its scope and students attempt to define the problem based on a generative topic that opens a field for the research. Students are encouraged to tune what they produce with their research questions. The learning process includes: observation, collaboration, fast-learning, visualization of ideas and prototyping. Assessment is based on prototyping and rebuilding which promote creative risk-taking. Design Thinking promotes the search for many possible options, not a single right answer. Therefore, it is successfully applicable to *any* course especially these with complex, multi-faceted and inter-disciplinary topics where innovation and creative problem solving is required (Interaction Design Foundation).

Practical employment of design thinking for social impact involves identifying the most pressing social problems, understanding the needs of the people, brainstorming solutions to social problems, testing and iterating on solutions to make them possibly most effective and finally building partnerships with other organizations to scale new solutions. The Hippo Water Roller is an example of a design-thinking solution to the problem of water scarcity in rural Africa. In 1994 two South African entrepreneurs, Pettie Petzer and Johan Jonker developed the roller, seeing how strenuous for women and children faced carrying water is. The roller is a large, round, barrel container that can carry up to 90 litres of water. It is pulled by hand, which makes it easier for women and children to transport water over long distances. It also has a filter cap to ensure that the water stored in the roller is pure. Petersen (2016) organized the lesson around the problem of designing the container. The comments on the innovation were the following: this is just a plastic barrel that holds water or even the Romans used pipes for water ... install a solar collector to run a water pump system to deliver water. These reactions revealed that students were not familiar with living conditions in rural Africa where there is no money for pipes and pumps and the rocky terrain is too difficult to lay plumbing and they could not imagine hardship of carrying water on a head for miles to survive. What makes the design thinking valuable is empathizing: "Until we walk a mile in someone else's shoes, we are blinded by our own culture, values and experience of what is a "breakthrough" that creates value" (Petersen, 2016).

5.4 Agile Methodology for Education, training and Security Projects

Projects in security education develop the desired skills to meet the requirements of today complex and unpredictable realities when the fast and collective response, self-organization determine the final results of the action. Projects are realized with the effort of the a cross-functional group of individuals that are collaborating towards a common goal (Malsam, 2024) by sharing knowledge, learning and building consensus. The relationship may vary depending on its strength but generally are defined by trust and commitment (Dietrich et al., 2010).

There are numerous methodologies to apply for project management including design thinking and agile. Yet, although both design thinking and agile methodology share principles of iteration, user-centred approach and collaboration, they are neither interchangeable nor mutually exclusive. Both methodologies along traditional ones may be applied without any conflict for security education and training as well as security operations, actions (Agile Alliance). The Agile Manifesto gives the foundation to 12 principles (Fig. 5.2).

5.4.1 PMO Global Institute Presents Examples when Agile Principles Are Applied in Project Process

- **Client Satisfaction:** An Agile project team delivers new features and updates, allowing the client to provide feedback.
- **Collaborative Approach:** Daily stand-up meetings allow developers, testers and product owners to coordinate their activities and address issues quickly.

12 Principles based on the Agile Manifesto			
1	The highest priority is to satisfy the customer through the early and continuous delivery of valuable software.	7	Working software is the primary measure of progress.
2	Welcome changing requirements, even late in development. Agile processes harness change for the customer's competitive advantage.	8	Agile processes promote sustainable development. The sponsors, developers, and users should be able to maintain a constant pace indefinitely.
3	Deliver working software frequently, from a couple of weeks to a couple of months, with a preference to the shorter timescale.	9	Continuous attention to technical excellence and good design enhances agility.
4	Business people and developers must work together daily throughout the project.	10	Simplicity–the art of maximizing the amount of work not done–is essential.
5	Build projects around motivated individuals. Give them the environment and support they need, and trust them to get the job done.	11	The best architectures, requirements, and designs emerge from self-organizing teams.
6	The most efficient and effective method of conveying information to and within a development team is face-to-face conversation.	12	At regular intervals, the team reflects on how to become more effective, then tunes and adjusts its behavior accordingly.

Fig. 5.2 Principles behind the Agile Manifesto

- **Frequent Delivery:** In a web development project, new website features are published every two weeks, so that visitors receive a steady stream of improvements.
- **Embracing Change:** An e-commerce company adjusts to new market trends by changing its website style and functionalities in response to customer feedback and market shift.
- **Sustainable Development:** An Agile team continually maintains a sustainable work tempo to avoid burnout and ensure the high-quality work (PMO Global Institute, 2023).

These principles can be very well applied in security projects. Humanitarian action of saving people's lives (evacuation, delivering food, etc.) from the war-torn areas will meet satisfaction not only of the saved but also those who provided support and help. These actions unquestionably require collaboration of different entities, bodies, agendas, services, stakeholders which needs to be coordinated and continually adjusted to uncertain and rapidly changing crisis or conflict situation. Updated feedback helps to embrace change and make right decisions.

Agile project management is a methodology used primarily in software development, yet it can be applied in other fields: security and education. Agile Favors flexibility and collaboration, incorporates customer feedback throughout the project life cycle. It takes an iterative approach to development, breaking down work into small, manageable cycles—"sprints" to focus on continuous improvement in the development of a product or service.

Core Competencies (OECD)	Cognitive Readiness	Transformative competencies (OECD)	Core Behavioural Competencies (UNDP)	Desing Thinking	Agile core competencies
Achieving results: analytical thinking , achievement focus, flexible thinking, drafting skills, managing resources, teamwork and team leadership	Situation awareness: the ability to perceive oneself in relation to the enemy and the environment)	Creating new value: sense of purpose, curiosity and an open mindset, critical thinking, creativity in finding different approaches to solving complex problems, collaboration, agility in trying out new ideas, ability to manage risks associated with these new ideas, adaptability skills	Achieve Results: focus on achieving quality results and impact	Conditioned inventiveness: creative thinking directed toward inventing.	Analytical thinking , problem-solving: analysing the situation; identifying the problem, causes, relationships and implications, opportunities; drawing from a range of resources, contexts and experiences; conceptualizing solutions; considering alternatives; and implementing the most appropriate action.
Interpersonal: building relationships (client focus, diplomatic sensitivity, influencing, negotiating, organisational knowledge)	Memory as an active, reconstructive process	Reconciling tensions and dilemmas: cognitive flexibility and perspective-taking, empathy and respect towards others, creativity and problem-solving skills, making complex and difficult decisions; a sense of resilience, tolerance for complexity and ambiguity, a sense of responsibility towards others.	Thinking Innovatively: seeing patterns and clarity outside boxes	Human-centred focus, client-directed and environment-centered concern, generating sustainable design	Building Interpersonal Relationships: displaying empathy to indicate understanding , creating an institutional climate in which people from different cultures feel welcome
Strategic: planning for the future (developing talent, organisational alignment, strategic networking, strategic thinking)	Transfer of training as the ability to apply what is learned in one performance context to another performance context	Taking responsibility: a strong moral compass, locus of control and sense of integrity, compassion and respect for others, critical thinking, a sense of self-awareness, self-regulation and reflective thinking building trust	Learn Continuously: pursuing self-development professionally and personally	Ability to visualize ideas in a range of media, bringing a common view to concepts otherwise imagined uniquely by everyone in a discussion.	Client Service and Support: making client needs a primary focus of actions

Core Competencies (OECD)	Core Competencies (OECD)	Cognitive Readiness	Transformative competencies (OECD)	Core Behavioural Competencies (UNDP)	Desing Thinking	Agile core competencies
	Metacognition: the executive functions of thought, pertaining to knowledge and regulation of one's cognitive processes and progress toward accepted goals			Adapt with Agility: adopting new ideas, working within new situations	Optimistic and proactive ways of working, controlling moods.	Communication: expressing ideas effectively in individual and group situations; listening effectively; sharing information, ideas and arguments; using appropriate non-verbal communication.
	Automaticity: processes that are performed rapidly, requiring few attentional resources			Act with Determination: courage, self-motivation and initiative	Accepting the view that solutions should be adaptive to fit the needs of users uniquely;	Planning and Organizing / Work Management: planning, scheduling, prioritizing and controlling activities; identifying, integrating and orchestrating resources (people, material, information, budget, and/or time) to accomplish goals.
	Problem solving: transforming goals into a plan of action by processes such as trial-and-error, proximity, fractionation, and knowledge-based referrals.			Engage and Partner: empathy and emotional intelligence, team-work	Predisposition toward multifunctionality: looking for multiple dividends from solutions to problems.	Teamwork / Collaboratio: Actively participating as a member of a team; contributing to and accepts the consensus; subordinating own needs to those of the team; developing and building cohesive team relations to produce required outputs;
	Decision-making: the selection of tactical and strategic plans			Enable Diversity and Inclusion (Treat all individuals with respect, consider ethical implications)	Systemic vision: treating problems as system problems with opportunities for sys-temic solutions involving mixes of hardware, software, procedures, policies, organizational concepts and whatever else is necessary to create a holistic solution.	

Core Competencies (OECD)	Cognitive Readiness	Transformative competencies (OECD)	Core Behavioural Competencies (UNDP)	Desing Thinking	Agile core competencies	Core Competencies (OECD)
	Leadership				Using language as a tool (visual, mathematical and verbal language)	
	Emotion: channelling and controlling them while performing complex tasks under the stress and confusion				Interpersonal skills	
					Facility for avoiding the necessity of choice: searching the competing alternatives for their essential characteristics and finding ways to reformulate them in a new configuration; the result is a solution that avoids the decision and combines the best of both possible choices	
					Self-governing practicality- a style of thinking that explores freely while maintaining a realistic appraisal of costs and functionality	
					Working systematically with qualitative information/ structured planning - a tool-kit of methods for a complete range of planning task: finding information, gaining insights from it, organizing it optimally for conceptualization, evaluating results and communicating a plan to the public and follow-on teams in the development process.	

Fig. 5.3 a. Generalized sets of competencies in the selected competency frameworks. Part 1. b. Generalized sets of competencies in the selected competency frameworks. Part 2. c. Generalized sets of competencies in the selected competency frameworks. Part 3

Agile project employs cross-functional teams who work closely with stakeholders to make adjustments as requirements evolve and formalizes feedback loops at the end of each sprint.

Agile project management is very well suited for a project which requirements are not fully specified from the outset or are likely to evolve over the project's duration. It is also best applicable when regular feedback is necessary for aligning project deliverables to customer expectations, when time-to-market is essential. A cross-functional, iterative approach enables teams to deliver product quicker and addressing issues as they arise, therefore Agile methodology is very well applicable for projects of high complexity (e.g. humanitarian aid, managing migration, evacuation of citizens from the war/crisis affected areas).

The situations when Agile may not be effective are if a customer is not certain about goals, the project manager or team lack experience or if they do not function well under pressure. Agile is less formal and more flexible processes, therefore it may not always be incorporated within large, traditional organizations which demonstrate high rigidity within processes, policies, or teams. It may similarly face problems when used with customers who have rigid processes or operating methods, or on projects subject to strict regulation and documentation, with definitely settled contracts and scope, or environments in which stakeholder feedback is unavailable or hardly accessible (Alexander, 2023).

5.4.2 *Real-Life Agile Application in Education*

There are several advantages for teaching practice that Agile methodology contributes. One of them are continuous and Incremental Learning in Shorter Phases. Semesters are well established and preferred learning period for most schools, yet it seems to be a prolonged timeframe with obvious drawbacks. Students often act slowly, are late or fail to complete the requirements of "a long-winding" curriculum. Feedback is also planned at the end of the semester where exams are organized to measure a student's performance through months of numerous courses. Students try their best to memorize, do the test successfully and then forget all about it. In Agile educational environment semesters are replaced with shorter two-week phases or *sprints* (as it is called in Scrum), as the preferred learning period. The sprints model is not about squeezing or reducing a full semester's learning backlog but prioritizing the backlog and plan the content to be taught to and learnt by students within a two-week sprint. An online learning platform would make learning more efficient and interactive. This sprint approach fosters teaching process by: speeding up feedback, enabling students to manage their learning workload easier and immediately applying what they have learned, whereas teachers can tailor their learning approach to the needs of the students as they know directly students' strengths and weaknesses. Other advantage of Agile for education is learning together. Traditional education focuses on individualized learning, whereas Agile promotes collaboration and interaction which enhances teaching/learning effectiveness. While working in teams, students manage their learning sprints by building communication, teamwork, social skills and improving their learning practice. In Agile making teams is a strictly defined process. Professors and students collaborate on the equal basis: they should be teams working on the same project—lectures. Professors do not lecture ex-cathedra any more but involve students in the learning process in class by giving students a choice of tasks to perform, taking individual responsibility and collaborating, giving feedback during the project. Students get first-hand experience of trial and error so that they get engaged (The Olive Press, 2023). Gamification is the next method well suited to the Agile education. Games introduce goals, interaction, feedback, problem solving, competition, narrative and fun learning environments—all these elements that can increase learner engagement and motivation. Yet, it is crucial to discuss the difference between gamification and game-based learning and their pedagogical values they can bring to both traditional and online courses.

Gamification "is the integration of game elements like point systems, leaderboards, badges, or other elements related to games into "conventional" learning activities in order to increase engagement and motivation" whereas game-based learning "involves designing learning activities so that game characteristics and game principles inhere within the learning activities themselves" e.g.: role-play a mock negotiation involving a labour dispute. Thus, gamification applies game elements to existing learning activities and game-based learning designs learning activities that are game-like (e.g. Kahoot) (University of Waterloo).

The scenarios for simulator Jaśmin may exemplify game-based approach. The scenario begins with the background information about the disappearance of a person (e.g. a notification, official note, description and photo of the wanted person). The first element of the activities is a short briefing, followed by the field activity of the Search and Rescue

Groups. A digital map in a system depicts the progress of crisis situation. The team leader defines the area for the search, assigns individual sections of the task and responsibilities to the particular students/soldiers. In the course of the actions, the responsibilities are changed or modified as the new information about the "wanted person" are collected. While searching the area students find a personal thing (its location appears on the map and the photo in the system) and as a result, the person. During the search, information between the groups is exchanged. Search and Rescue Groups maintain a permanent communication channel with the police's command post via chat and voice, depicting video image, photos, chat, data and GPS coordinates from BSP, GoPro and thermal cameras, via Video Streaming Server. During the task all members of groups are equipped with SZK *Jaśmin* elements (tablets and smartphones with installed SZK *Jaśmin* software) and monitored. These elements help to build situational awareness and develop integration and cooperation of various services within one functional system with data provided. When task is completed, the results are summarized and discussed with *the After-Action Review* functionality (ZBiAM, 2022). (The examples of the real use of Jaśmin were described in Chapter 2).

The next Agile feature for education is teaching together or Paired Teaching concept. In the software the concept is called Pair Programming—two developers take turns in trying to code and develop a feature, doing peer reviews and making an implementation. These procedures may be well applied in teaching—teachers pair up to design and implement lesson plans together. One teacher is a guest lecturer at another teacher's class. Then, they exchange notes about their experiences and for example perform their own teaching sprints with modified goals and deliverables. Agile help also prioritize Better Learning Experiences as it makes headmasters and teachers collaborate and come up with initiatives. Teachers who want to make learning experiences for their students more engaging may need to experiment until they reach a satisfactory result. The last feature Agile offers is constant reflection at all levels. Agile teachers' teams should regularly reflect on their performance and experience to improve their teaching styles and process in the next iteration. The same applies to students. Working in teams, students can perform their reflection sessions—"retrospectives" to identify the areas for improvements (Burstein; Boiser, 2024).

5.5 Competencies Developed—Discussion and Conclusions

As stated above both presented methodologies are suitable for educational and security project, yet there are several differences between them. Design thinking focuses on exploring and defining the right problem and solution, whereas agile—on efficiently executing and delivering a product. The design thinking originates from design and builds on multiple disciplines: psychology, systems thinking and business strategy. The agile draws from software development and disciplines such as manufacturing and project management. DT aims to define the problem, test and pick a solution and agile delivers that solution and continuously iterate on the live product. Thus, in DT an idea or solution is usually a prototype (may be intangible) and agile provides a tangible, working product for an end user (Interaction Design Foundation, 2016). Both methodologies are powerful tools that enable security services, educators and other stakeholders to adapt and innovate to survive in a dynamic crisis situation. The adaptive capability results from development of appropriate competencies.[2] The proposed sets of competencies that allows to operate in crisis situation are:

- Transformative competencies for 2030 understood as

[…] the types of knowledge, skills, attitudes and values students need to transform society and shape the future for better lives. These have been identified as creating new value, reconciling tensions and dilemmas and taking responsibility" (OECD, 2019, p. 63).

- Cognitive readiness:

[…] the mental preparation (including skills, knowledge, abilities, motivations, and personal dispositions) an individual needs to establish and sustain competent performance in the complex and unpredictable environment of modern military operations (Morrison & Fletcher, 2002, p.

[2] Adaptability and competencies are also discussed in: D. Kaźmierczak (2021). Competencies for viable subsistence", EEPublishing.

ES-1) Cognitive readiness may be understood as a combination of three basic abilities to: recognize patterns in chaotic situations, modify problem solutions associated with these patterns appropriate to the situation, implement action plans drawn on these solutions.

- Core competencies:

> […] the capabilities that are important across all jobs and that we believe collectively contribute to the OECD's overall success. At the same time, the importance of Core Competencies may vary according to the specific job duties and requirements. The OECD Competency Framework displays fifteen Core Competencies grouped into three clusters […] the delivery-related competencies […] interpersonal competencies […] strategic competencies (OECD, 2014, p. 4).

- Core Behavioural Competencies

> […] capture the attitudes and behaviours expected of every individual working in the organisation. They help answer the question about the kind of people UNDP wants and needs to have among its workforce to be able to fulfil its mission. There are seven competencies that are relevant to everyone at UNDP: […] achieve results, think innovatively, learn continuously, adapt with agility, act with determination, engage and partner, enable diversity (UNDP, p.1).

- Design thinking is more than a process it requires particular mindset:

> Design thinking mindsets are how individuals think, feel and express themselves during design thinking activities. It includes people's expectations and orientations during a design project (Interaction Desing Foundation).

The key mindsets that ensure a successful design thinking process are: being empathetic, collaborative, curious, optimistic, embracing ambiguity and diversity, able to reframe, making ideas tangible and make experiments and learn from them.

- Agile core competencies

> [...] required for all positions at organizations: Building interpersonal relationships; Analytical thinking /Problem solving; Client service and support; communication, planning and organizing/work management; Teamwork/collaboration; organization awareness (Agile Leaders Project Management Services, p.8).

The list of competencies common to all the sets quoted above includes: problem solving, creative thinking towards inventing, building interpersonal relations, empathy, client-centred solutions, adaptability/agility, self-governing and work management. Table 6.1 presents generalized sets of competencies in the selected six frameworks.

The latest trends shaped by the changing and complex realities show that being agile is required but not sufficient as there is growing demand for cross-skilled individuals and consequently skill-based learning. Human skills like communication, collaboration, problem-solving, and creativity so well developed and mastered through the design thinking and agile methodologies, are becoming even more important than functional skills. Individuals are expected to have multiple deep skill capabilities, not just functional skills to perform a particular role (Business Agility Institute, Scrum Alliance).

References

What Is Connectivism Learning Theory and How Can You Apply It in Learning and Development? (n.d.) https://360learning.com/guide/learning-theories/connectivism-learning-theory/

Agile Alliance. *AGILE ESSENTIALS. The 12 Principles behind the Agile Manifesto.* https://www.agilealliance.org/agile101/12-principles-behind-the-agile-manifesto/

Skinner, B. F. (1976). *About Behaviorism*. Vintage Books. https://gsi.berkeley.edu/gsi-guide-contents/learning-theory-research/behaviorism/

Soutter, Will. (2023, June 17). Nanotechnology in the military. AZoNano. Retrieved on November 16, 2024 from https://www.azonano.com/article.aspx?ArticleID=3028.

Stevens, E. (2024, Sept. 10). 7 fundamental user experience (UX) design principles all designers should know. https://www.uxdesigninstitute.com/blog/ux-design-principles/

Stobierski, T. (2022). *5 design thinking skills for business professionals*. Business Insights. Harvard Business School. https://online.hbs.edu/blog/post/design-thinking-skills

Structural Learning.com. (2022, Dec. 9). Cognitivism learning theories: A teachers guide. https://www.structural-learning.com/post/cognitivism-learning-theories

The Olive Press. (2023, June 1). Agile Project Management in the education sector. https://www.theolivepress.es/spain-news/2023/06/01/agile-project-management-in-the-education-sector/

UNDP (n.d.). People for 2030—Competency framework–Core behavioural competencies.

United Nation Economic and Social Council. E/C.16/2006/4 (2006). Definition of basic concepts and terminologies in governance and public administration.

University of Waterloo (n.d.). Centre for Teaching Excellence. Gamification and game-based learning. https://uwaterloo.ca/centre-for-teaching-excellence/catalogs/tip-sheets/gamification-and-game-based-learning

WGU. (2020, May 28). What is the behavioural learning theory?. https://www.wgu.edu/blog/what-behavioral-learning-theory2005.html#close

World Economic Forum Geneva. (2018). *The global risks report 2018* (13th ed.). https://iapss.org/2014/04/22/the-five-main-categories-of-risks-at-the-global-level/

ZBiAM. (2022, June 2). Jaśmin also for crisis management. https://zbiam.pl/jasmin-tez-dla-zarzadzania-kryzysowego/

Zostań Żołnierzem (n.d.). Wojsko Polskie. https://www.wojsko-polskie.pl/zostanzolnierzem/legia-akademicka/

Conclusions–Reflections on the Future of Cities

Our research on city as a dynamic and complex ecosystem, focused on how the city growths and what are its strengths and vulnerabilities when the crisis hits?

Undoubtedly, technology is a distinct driving factor for urban development bringing about change in both quality of life and the mindset. The mindset is well reflected in the smart city concept which builds on: citizen-centric approach, data-driven decision-making, sustainability, innovation and entrepreneurship, collaboration and resilience. Rather than addressing challenges separately, substantial economic and environmental gains can be made from adopting more integrated approaches. However, the concept is questioned as it generates the gains only for the privileged groups separated from the unprivileged poverty, inequality, unemployment, illiteracy, corruption, violence, lack of sanitary, educational and health care systems (Gonella, 2019). Building city resilience to adverse situations, crises, natural disasters, conflicts and social upheavals is even more a challenge in the face of its expansion into city-regions – an urban area at metropolitan and larger scales (Neuman & Hull, 2009). The growth of the cities on one hand, poses the challenge to city affordability (migration, influx of students, the tech elite and cheap work force) but on another hand, it is to be a solution to occurring problems and systemic weaknesses. The recent exemplification of the growing city–region is the concept of the new city in Silesia, which will exceed the borders of the present towns. The Upper Silesian-Basin Metropolis (GZM) with 41 cities

S. Kowalkowski et al., *Cities in Crisis*,
https://doi.org/10.1007/978-3-032-08843-7

and municipalities in the area of 2.500 square kilometres and population of 2.1. million residents, is the first metropolis in Poland. The main goal behind creating a new city is to improve the quality of life of residents by exploiting the potential of individual localities within one organism and enhance its position in the country and Europe. This change does not involve liquidation of existing local governments but introducing a new division of competences and create a metropolitan municipality that would deal with tasks of a supra-local nature. (Kędzierski, 2024). The newly created strategies for city-region management and development demonstrate the agility (adaptive capabilities) of leaders and the ecosystem itself, which are very crucial in the face of emerging crises, potential conflicts and wars.

Cities have always been and will be strategic elements in the national defence plans; controlling a city which is a core hub of people, power, business, social and cultural institutions and a realization of national identities is strategically critical. The concept of urban combat exploits both physical (terrain and infrastructure) and social assets and offers advantage to the defender mainly by mitigating the technological superiority of a more powerful opponent. City as a complex and multi-dimensional ecosystem requires parallelly designed form of action – multidomain operations–both concepts intersect in the point of layers and convergence, concept of winning without fighting (applying soft power, or blurring the line between “below conflict” and conflict) and extension and delineation of the battlefield by use of high-tech for cyberattacks, information campaigns, terrorist actions and traditional kinetic actions. Flexible use of all assets will allow NATO MDO to conduct high-intensity combat, law enforcement, humanitarian operations and military advisory operations (Wolfel et. al., 2021).

It is also crucial to focus also on limitations of urban warfare which include: variety of buildings and objects, especially those inhabited by civilians; variety and uniqueness of each city ecosystem that requires every single planning and completing the military operations in the urbanized area with different procedures, military structures, specialized weapons; urban operations become long and devastating when the offensive forces try to break the defendants’ will by bombing and artillery shelling.

As a consequence, the human impact is dramatic. The damaged infrastructure entails disruptions in providing basic services: food, water, shelter, sanitation, electricity and health care. The situation is even worse when humanitarian aid is blocked or the conflict protracts. Victims forced to displace may be caught in crossfire, mistreated by parties to armed conflict, treated as live shields and separated from their family members.

Although the military instructions and doctrines of most armies strongly recommends omitting or isolating urbanized areas, in practice this recommendation contradicts the basic laws of war which prove that capturing the capital city or political centre may determine the outcome of the campaign.

The serious challenge posed to disaster risk management systems by military conflicts, protracted crises and compound disasters requires integrated, cross-sectoral response.

This response should not focus on "militarization and luxury redevelopment" as keys to the post-crisis "resilient city", as Miriam Greenberg notices (2017) but rather on cities informal settlements and refugee camps to strive for environmental and economic justice, racial and ethnic inclusion and civil society engagement. Such an approach requires addressing cities' "temporality and spatiality, politics and multiple potentialities".

The model of Disaster Risk Management integrated from four interconnected dimensions: sectoral, spatial/hierarchical, temporal and externally with other cross-cutting societal challenges is the solution that could optimize the humanitarian aid. In the integrated DRM system people, businesses, systems, technology, processes and projects are brought together to achieve operational unity and complete tasks effectively across the all supply chains of the city ecosystem.

The well-known integration of institutional assets, coordination between the state and the third sector in the provision of goods and services viewed as an expression of and urban citizenship have been challenged by pollicises. The response to complex needs requires new forms of solidarity - hybrid frameworks including as a combination of voluntary non-profit activities on one hand and social movements and political activism on the other, negotiated and coproduced in broader societal self-organizing spaces. They are expected to make a difference especially in the area of health, migration or housing–the critical issues for cities in crisis and post-crisis conditions (Pajnik et. al., 2025).

The multidimensional integration is a process. It cannot be achieved with one decision or action and needs to be seen as a capability to adapt by agile reconfiguration of means and forces. The adaptive capability is inscribed in the concepts of design thinking and agile methodologies that are applicable not only in business and social context to make the difference but also in education and training to enhance the desired competencies of the security services and society.

These generalized considerations on the present and future condition of cities which face variety of challenges will certainly be further confronted with the dynamic realities of the today's world, but we believe they contribute to discussion on effective preparation, response and recovery from the crises that are to come.

References

Kędzierski, R. (2024, Nov.19). *Polska będzie miała zupełnie nowe miasto? Ma być największe w kraju*. https://www.money.pl/gospodarka/polska-bedzie-miala-zupelnie-nowe-miasto-ma-byc-najwieksze-w-kraju-7080881342728800a.html

Siebel, W. (2018). Urbanity. Akademie für Raumforschung und Landesplanung. https://nbn-resolving.org/urn:nbn:de:0156-559926129.

Gonella, F. (2019). The Smart Narrative of a Smart City. *Front. Sustain. Cities, 1*, 9. https://doi.org/10.3389/frsc.2019.00009

Greenberg, M. (2017, Feb.). The City at War. *International Journal of Urban And Regional Research*. https://www.ijurr.org/spotlight-on/the-city-at-war-reflections-on-beirut-brussels-and-beyond/the-city-at-war/

Neuman, M., & Hull, A. (2009). The Futures of the City Region. *Regional Studies, 43*(6), 777–787. https://doi.org/10.1080/00343400903037511

Pajnik. M. et al. (eds.), Framing Solidarities in Times of Multiple Crises, Palgrave Studies in European Political Sociology, https://doi.org/10.1007/978-3-031-82563-7_1

State of Green. (2020). SMART CITIES. Creating liveable, sustainable and prosperous societies Version 2.0.

Wolfel, R.L., Richmond, A., Ridgeway, J. (2021). Dense Urban Environments The Crucible of Multi-Domain Operations. *Military Review*, January-February 2021, 22–32. https://www.tree-house.co.uk/blog/what-is-the-difference-between-integration-and-project-management

S. Kowalkowski et al., *Cities in Crisis*,
https://doi.org/10.1007/978-3-032-08843-7

Index

S. Kowalkowski et al., *Cities in Crisis*,
https://doi.org/10.1007/978-3-032-08843-7